Additive Manufacturing Technologies and Applications

Special Issue Editors

Salvatore Brischetto

Paolo Maggiore

Carlo Giovanni Ferro

MDPI • Basel • Beijing • Wuhan • Barcelona • Belgrade

MDPI

Special Issue Editors

Salvatore Brischetto
Politecnico di Torino
Italy

Paolo Maggiore
Politecnico di Torino
Italy

Carlo Giovanni Ferro
Politecnico di Torino
Italy

Editorial Office
MDPI AG
St. Alban-Anlage 66
Basel, Switzerland

This edition is a reprint of the Special Issue published online in the open access journal *Technologies* (ISSN 2227-7080) from in 2017 (available at: http://www.mdpi.com/journal/technologies/special_issues/additive_manufacturing).

For citation purposes, cite each article independently as indicated on the article page online and as indicated below:

Author 1; Author 2. Article title. *Journal Name* **Year**, *Article number*, page range.

First Edition 2017

ISBN 978-3-03842-548-9 (Pbk)
ISBN 978-3-03842-549-6 (PDF)

Table of Contents

About the Special Issue Editors

Salvatore Brischetto, assistant professor at the Department of Mechanical and Aerospace Engineering of Politecnico di Torino since 2010, received his PhD in Aerospace Engineering (Politecnico di Torino) and in Mechanics (Université Paris Ouest–Nanterre La Défense) in 2009. His main research topics are smart composite and FGM structures, multifield problems, shell 3D and 2D numerical and exact solutions, additive manufacturing and UAVs. He is the author of more than 100 articles, more than 60 of which have been published in international journals. He is a committee member for several international journals and one book series, and member of the "Shell Buckling People" web-site. Since the academic year 2013/2014 he has been adjunct professor for the course "Aeronautic law and human factors and safety". He is co-founder and co-chair of the research group "ASTRA: Additive manufacturing for Systems and sTRuctures in Aerospace".

Paolo Maggiore is associate professor at the Mechanical and Aerospace Engineering Department of Politecnico di Torino, which he joined in 1992, and where he teaches aerospace general systems engineering. Currently, his research interests range from hydrogen fuel cell powered airplanes and UAVs, and health monitoring of flight controls, to additive manufacturing technologies, and multi-disciplinary design optimization of aerospace systems design.

Carlo Giovanni Ferro is a PhD Student at the Mechanical and Aerospace Engineering Department of Politecnico di Torino. Together with Professors Maggiore and Brischetto, he is a founder of the Astra Research Group. He has past experience in the Additive Manufacturing environment both in industrial (Ge Aviation) and Research Centres (CERN). Currently, his research is devoted to lattice structures for multipurpose panels for aerospace utilization.

technologies

MDPI

Editorial

Special Issue on "Additive Manufacturing Technologies and Applications"

Salvatore Brischetto * , **Paolo Maggiore and Carlo Giovanni Ferro**

ASTRA (Additive Manufacturing for Systems and Structures in Aerospace) Group, Politecnico di Torino, Department of Mechanical and Aerospace Engineering, Corso Duca degli Abruzzi 24, 10129 Torino, Italy; paolo.maggiore@polito.it (P.M.); carlo.ferro@polito.it (C.G.F.)
* Correspondence: salvatore.brischetto@polito.it

Received: 8 September 2017; Accepted: 11 September 2017; Published: 12 September 2017

Additive Manufacturing (AM) is a well-known technology, first patented in 1984 by the French scientist Alain Le Mehaute. Its distinctive feature is the addition of material with different methods (e.g., powder or wire) in place of the subtraction of material from a raw part. AM has been widely introduced in the preliminary and conceptual design phase, thanks to its reduced production costs and realization time for a prototype. In the last two decades, this technique has also been considered for low-scale mass production due to some advantages. This method allows the construction of so-called evolutionary shapes: structures of complex design that are impossible or difficult to build via traditional milling or machining. Evolutionary shapes are usually the result of a topological optimization. For these reasons, important mass savings or increases in structure mechanical properties are obtained using AM. The present Special Issue proposes articles in the area of Additive Manufacturing with particular attention to the different employed technologies and the several possible applications. The main investigated technologies are the Selective Laser Sintering (SLS) and the Fused Deposition Modelling (FDM). These methodologies, combined with the Computer Aided Design (CAD), provide important advantages. Numerical, analytical and experimental knowledge and models are proposed to exploit the potential advantages given by 3D printing for the production of modern systems and structures in aerospace, mechanical, civil and biomedical engineering fields.

This Special Issue of *Technologies* comprises 11 selected papers about different additive manufacturing methodologies and related applications and studies. The first paper by Petersen et al. analyses the significant impact in the near future of Do-It-Yourself (DIY) manufacturing via 3D printing on the toy and game markets. The second paper by Laureto and Pearce describes a manufacturing technology that allows a constrained set of metal–polymer composite components. The main conclusion is that an open source software and hardware tool chain can provide low-cost industrial manufacturing of complex metal–polymer composite-based products. Ferro et al. propose a multi-functional panel concept that integrates the anti-icing system directly inside the primary structure. The core of the sandwich includes trabecular non-stochastic cells that allow the presence of a heat exchanger directly embedded in the leading edge. This solution is easily produced in a single-piece component using Additive Manufacturing (AM) technology without the need of joints, gluing or welding. A preliminary investigation of the mechanical properties of the core produced via the Selective Laser Melting (SLM) method is proposed. Mazzucato et al. propose the monitoring of a new deposition nozzle solution for Direct Energy Deposition (DED) systems through a simulation–experimental comparison. Preliminary tests are carried out by varying powder, carrier and shielding mass flow, demonstrating that the last parameter has a significant influence on the powder distribution and powder flow geometry. Vora et al. show the creation of an Al339 alloy from compositionally distinct powder blends. The in-situ alloying of this material and the Anchorless Selective Laser Melting (ASLM) processing conditions allow components to be built in a stress-relieved state, enabling the manufacture of overhanging and unsupported features. This novel method, known as ASLM, maintains processed material within a stress-relieved state throughout the duration of a build. Kutzer and DeVries discuss a new methodology that applies material onto or around existing surfaces with multilayer and thick features. The main novelties of the paper are the derivation of deposition paths giving a prescribed set of layers; the design, characterization and control of a proof-of-concept testbed; and the derivation and application of time evolving trajectories subjected to the material deposition constraints and mechanical constraints of the testbed. Results show the feasibility of conformal material deposition with multilayer and thick features. Brischetto et al. propose Fused Deposition Modelling (FDM) characterization in order to apply this technology in the construction of aeronautical structural parts when stresses are not excessive. A statistical characterization of the mechanical properties of ABS (Acrylonitrile Butadiene Styrene) specimens during compression tests is proposed. A capability analysis is also used as a reference method to evaluate the boundaries of acceptance for both mechanical and dimensional performances. The statistical characterization and the capability analysis are proposed in an extensive form in order to validate a general method that will be used for further tests in a wider context. Ilie et al. show that the layer-by-layer building methodology, used within the powder bed process of Selective Laser Melting (SLM), facilitates control over the degree of melting achieved at each layer. This control can be used to manipulate levels of porosity within each layer, affecting resultant mechanical properties. The results indicate that there is potential to use SLM for customising mechanical performances over the cross-section of a component. Prashanth et al. study the properties of five different metals/alloys fabricated by means of SLM. The results show that SLM is a reliable fabrication method to produce metallic materials with consistent and reproducible properties. Petersen and Pearce study a representative model for the potential future of 3D printing in the average American household by employing a printer operator who is relatively unfamiliar with 3D printing and 3D design files of common items normally purchased by the average consumer. Twenty-six items are printed in thermoplastic and a cost analysis is performed through comparison to comparable and commercially available products at a low and high price range. The paper by Patterson et al. analyses SLM and the Direct Metal Laser Sintering (DMLS) processes. SLM/DMLS can produce full-density metal parts from difficult materials, but it tends to suffer from severe residual stresses introduced during processing. This feature limits the usefulness and applicability of the process, particularly in the fabrication of parts with delicate overhanging and protruding features. The purpose of this study is to examine the current status and progress made toward understanding and eliminating the problem in overhanging and protruding structures.

The articles published in this Special Issue present only some of the most important topics about additive manufacturing technologies and applications. However, the selected papers offer significant studies and promising methodologies.

Acknowledgments: The Guest Editors would like to thank all the authors for their invaluable contributions and the anonymous reviewers for their fundamental suggestions and comments.

Conflicts of Interest: The authors declare no conflict of interest.

technologies

MDPI

Article

Impact of DIY Home Manufacturing with 3D Printing on the Toy and Game Market

Emily E. Petersen [1], Romain W. Kidd [2] and Joshua M. Pearce [1,3,*] (iD)

[1]   Department of Material Science and Engineering, Michigan Technological University,
      Houghton, MI 49931, USA; eepeters@mtu.edu
[2]   MyMiniFactory, London E1 2JA, UK; romain@myminifactory.com
[3]   Department of Electrical and Computer Engineering, Michigan Technological University,
      Houghton, MI 49931, USA
*   Correspondence: pearce@mtu.edu; Tel.: +1-906-487-1466

Received: 16 May 2017; Accepted: 6 July 2017; Published: 20 July 2017

Abstract: The 2020 toy and game market is projected to be US$135 billion. To determine if 3D printing could affect these markets if consumers offset purchases by 3D printing free designs, this study investigates the 100 most popular downloaded designs at MyMiniFactory in a month. Savings are quantified for using a Lulzbot Mini 3D printer and three filament types: commercial filament, pellet-extruded filament, and post-consumer waste converted to filament with a recyclebot. Case studies probed the quality of: (1) six common complex toys; (2) Lego blocks; and (3) the customizability of open source board games. All filaments analyzed saved the user over 75% of the cost of commercially available true alternative toys and over 90% for recyclebot filament. Overall, these results indicate a single 3D printing repository among dozens is saving consumers well over $60 million/year in offset purchases. The most common savings fell by 40%–90% in total savings, which came with the ability to make novel toys and games. The results of this study show consumers can generate higher value items for less money using the open source distributed manufacturing paradigm. It appears clear that consumer do-it-yourself (DIY) manufacturing is set to have a significant impact on the toy and game markets in the future.

Keywords: toy industry; additive manufacturing; 3D printing; consumer; economics; open-source

1. Introduction

After 20 years of legal intellectual monopoly, the fused filament fabrication (FFF) technology of additive manufacturing (AM), where a single layer of polymer is deposited after another, was unshackled by the open source release of the self-REPlicating RAPid prototyper 3D printer (RepRap) [1–3]. This open source hardware approach [4] led to a rapid technical evolution, which resulted in aggressive cost declines and the desktop 3D printer market emerged [5], dominated by various RepRap derivative machines [6,7]. Early adopters of these 3D printers were largely used for prototyping and the maker community, but this has morphed into peer production [8]. Digital peer-production with RepRaps found an eager audience among scientists to develop experimental tools [9–12]. In addition, teachers adopted the technology looking for high-quality educational experiences for their students [13–17] as well as those looking for economic sustainable development with appropriate technologies [18,19]. Sales of desktop 3D printers, however, are now moving towards the mass consumer market [6].

A pair of recent studies indicate that 3D printing technology is lucrative to adopt for average consumers. In the first study, the purchase of US$500 components of a RepRap were justified by printing a handful of consumer products [20]. However, not all consumers are technically sophisticated makers able to build such a complex mechatronic technology alone, so a second study [21] looked at the use of an open source fully-assembled 3D printer (Lulzbot Mini, which retails for US$1250). The costs of printing 26 free designs were compared with purchasing commercial equivalents and the study found that producing consumers would earn nearly a 1000% return on investment (ROI) from the purchase of a 3D printer over a printer lifetime of five years printing out only one product a week [21]. In addition, it appeared that consumers had already offset over $4 million in purchases only from those random 26 products, which indicates that as 3D printer use in the home becomes more widespread, distributed manufacturing with open source designs could begin to have a significant macroeconomic impact [21]. There is significant skepticism of this potential [22] as the Foxconn CEO famously referred to 3D printers as "a gimmick" [23] and popular representations of 3D printers used only for toys in the home. However, the toy market is substantial and is not so easily dismissed, with the U.S. average spending per child on toys being $371/year [24] resulting in a U.S. market of more than US$10 billion/year [25] in 2013. In 2016, the NPD Group's Retail Tracking Service noted that the U.S. toy market had grown to $20.36 billion [26] and the global toys and games market is projected to reach US$135 billion by 2020 [27]. What are the effects on these markets if consumers are offsetting purchases of products like toys with 3D printing and free designs now?

To probe the potential economic impact of home use of 3D printing technology, this study closely investigates consumer use of a popular free website (MyMiniFactory) for 3D printable products. An economic analysis is performed from the perspective of users producing toys for themselves in their own homes. Specifically, the economic savings of the top 100 most popular designs as indicated by downloads (not views) on MyMiniFactory are quantified for January 2017. These savings are quantified using the sliced mass of filament and electricity consumption of a Lulzbot Mini 3D printer, the U.S. average electricity rate and three prices of filament: (1) the most popular filament sold on Amazon; (2) the use of a plastic extruder to make filament from commercial plastic pellets; and (3) the use of a home recyclebot to convert waste post-consumer waste into filament. The type of product is also evaluated and, because of the preponderance of products that can be classified as toys and games, three detailed case studies are undertaken. First, six common toys with equivalent products are evaluated in detail for functionality and value. Next, as Lego currently dominates the toy market [28] an economic evaluation is run using 3D printers only as Lego-compatible block factories. Finally, the costs and customizability are evaluated for an open source board game. Overall, the results are discussed in the context of distributed manufacturing in consumer homes and economics of do-it-yourself (DIY) production.

2. Materials and Methods

2.1. Download Value Quantification

A selection of items comprised of the 100 most downloaded files in January 2017 from MyMiniFactory, a repository for free 3D printable objects, was selected for analysis. The design and number of downloads (N_d) is shown in Table 1.

Table 1. Most popular designs downloaded on MyMiniFactory in January 2017.

Open Source 3D Printed Design	URL (www.myminifactory.com/object)	N_d
Pokemon Go aimer	pokeball-aimer-pokemon-go-23009	20583
Clash of Clans barbarian	barbarian-lv-1-clash-of-clans-858	8107
Voltron figure	voltron-defender-of-the-universe-22430	3881
Overwatch tracer gun	tracer-gun-overwatch-19011	3602
Overwatch reaper mask	reaper-mask-19004	3457
Overwatch McCree revolver	updated-mccree-revolver-by-jeff-lagant-not-me-19543	3303
Star Wars AT-AT	detailed-at-at-17606	3218
Last Word Destiny Hand Cannon	destiny-last-word-exotic-hand-cannon-6546	3140
Overwatch D.VA Light Gun	d-va-s-light-gun-18920	3135
Overwatch Reaper Shotgun	reaper-s-hellfire-shotguns-overwatch-19096	2943
Batman cowl	batman-cowl-20596	2926
Destiny Hawkmoon gun	destiny-hawkmoon-exotic-hand-cannon-6545	2885

Table 1. *Cont.*

Open Source 3D Printed Design	URL (www.myminifactory.com/object)	N_d
Destiny thorn gun	thorn-from-destiny-4494	2846
Star Wars VII Storm Trooper Helmet	star-wars-storm-trooper-vii-fully-wearable-helmet-12992	2588
Kylo Ren helmet	jj-industries-kylo-ren-helmet-14106	2451
Wall outlet shelf	wall-outlet-shelf-6382	2350
Kylo Ren lightsaber	kylo-ren-s-lightsaber-star-wars-6791	2347
Blade Runner blaster	deckards-blaster-blade-runner-5694	2337
Fallout 3 T45-d helmet	fallout-3-t45-d-power-armour-helmet-15253	2318
Venus de Milo figurine	venus-de-milo-at-louvre-paris-1657	2073
Warcraft Frostmourne sword	frostmourne-from-warcraft-4156	1999
3DR Iris+ quadcopter	3drobotics-iris-19615	1871
Pieta figurine	pieta-in-st-peter-s-basilica-vatican-3796	1862
P08 Luger gun	p08-luger-functional-assembly-5545	1798
Game of Thrones iron throne	game-of-thrones-iron-throne-1945	1786
The Thinker figurine	the-thinker-at-the-muse-rodin-france-2127	1755
Overwatch McCree Peacemaker gun	mccree-peacemaker-overwatch-19152	1723
Fallout 4 Pipboy 3000 MkIV	fallout-4-pipboy-3000-mkiv-16884	1705
Articulated lamp	articulated-lamp-6790	1704
Overwatch throwing star	genji-s-shuriken-18918	1695
Strong bolt	support-free-bolt-1281	1571
Mazigner Z Super Robot	mazigner-z-super-robot-24533	1570
Secret shelf	secret-shelf-3504	1564
Sombra pistol	sombra-s-machine-pistol-25186	1544
Han Solo blaster	han-solo-s-blaster-star-wars-1546	1520
Overwatch D.VA headset	d-va-headset-22077	1502
Overwatch Mercy staff	mercy-s-staff-22079	1484
Overwatch Mercy blaster	mercy-s-caduceus-blaster-18912	1481
Han Solo blaster	hans-solo-blaster-2488	1461
Gears of War Chainsaw gun	gears-of-war-lancer-chainsaw-gun-11478	1437
Destiny Duke MK.44 gun	duke-mk-44-hand-cannon-from-destiny-2140	1418
Fallout 4 Laser pistol	fallout-4-laser-pistol-18978	1417
Overwatch loot box	overwatch-loot-box-21670	1412
Cat at British Museum	gayer-anderson-cat-at-the-british-museum-london-4010	1364
Halo 5 assault rifle	halo-5-assault-rifle-11734	1334
Portal gun	portal-gun-18342	1328
Destiny hawkmoon gun	hawkmoon-from-destiny-full-scale-and-moving-6863	1327
Clash of Clans figurine	p-e-k-k-a-lv-1-clash-of-clans-857	1314
Harry Potter elder wand	dumbledore-s-elder-wand-2057	1299
Groot flower pot	baby-groot-flower-pot-gardens-of-the-galaxy-2-26442	1269
Melted Darth Vader mask	darth-vader-melted-mask-6685	1264
Fallout 4 10mm pistol	fallout-4-10mm-pistol-10475	1264
Starcraft Kerrigan statue	starcraft-kerrigan-statue-10432	1263
Destiny ghost	destiny-ghost-6038	1248
Lich king figurine	the-lich-king-6174	1212
Micro game bit	micro-bit-game-bit-13822	1199
Ant Man helmet	ant-man-mask-wearable-5322	1193
Game of Thrones House emblem	house-stark-game-of-thrones-1154	1182
Planetary gears	planetary-gears-1557	1163
Clash of Clans king figure	barbarian-king-clash-of-clans-871	1160
Tooth toothbrush holder	the-big-tooth-2-0-5759	1141
Halo 3 ODST helmet	halo-3-odst-helmet-wearable-cosplay-17614	1134
Fallout 4 protectron figure	fallout-4-protectron-action-figure-15585	1129
BFG Doom	bfg-21395	1122
Discobolus figurine	discobolus-at-the-british-museum-london-7896	1118
Anonymous mask	guy-fawkes-anonymous-mask-2582	1092
Guardian of the Galaxy Star Lord mask	guardians-of-the-galaxy-star-lord-s-mask-version-2-3045	1091
Pokeball	pokeball-high-detail-version-23506	1064
Witcher 3 wall plaque	the-witcher-3-wall-plaque-8882	1063
Overwatch widowmaker rifle	overwatch-widowmaker-sniper-rifle-21702	1029
Fallout 4 combat rifle and shotgun	fallout-4-combat-rifle-and-combat-shotgun-18428	1025
Michelangelo's David	michelangelo-s-david-in-florence-italy-2052	1023
Destiny ghost	ghost-destiny-2396	1022
Triceratops skull	triceratops-skull-in-colorado-usa-6225	1019
Skull ring	skull-ring-20782	1014
Star Wars NN-44 Rey's Blaster	star-wars-nl-44-reys-blaster-17422	1012
Joker mask	joker-mask-9743	1010
Fork/spoon support for disability	fork-and-spoon-support-for-person-with-disabilities-5480	1007
BFG Doom	bfg-parts-19092	1006
Gryffindor coat of arms	gryffindor-coat-of-arms-wall-desk-display-harry-potter-11834	994
Destiny Gjallahorn	gjallarhorn-2-0-destiny-19160	985
Pokemon figurines	low-poly-pokemon-collection-15905	983
Rubik's cube	3d-printable-rubik-s-cube-9734	980
Destiny bad juju pulse rifle	destiny-s-bad-juju-exotic-pulse-rifle-6618	977
Fallout 4 Kellogg's pistol	kellogg-s-pistol-fallout-4-21556	975
Dobby the Elf figurine	dobby-from-harry-potter-full-model-3294	970
Statue of Liberty figurine	statue-of-liberty-in-manhatten-new-york-2077	960
Nutcracker	nutcracker-v2-4361	946
Minions figurines	minion-movie-trio-10140	940
Destiny sleeper simulant	the-sleeper-simulant-from-destiny-14769	930

Table 1. *Cont.*

Open Source 3D Printed Design	URL (www.myminifactory.com/object)	N_d
Frozen Elsa figurine	elsa-from-disney-s-frozen-6573	919
Star Wars storm trooper rifle	blaster-rifle-star-wars-storm-trooper-12097	914
Vitruvian Man scuplture	the-vetruvian-man-sculpture-at-belgrave-square-london-1669	908
Robocop ED209 figure	ed209-from-robocop-5090	904
Harry Potter wand	harry-potter-s-wand-10391	898
Cable guards	icableguards-21235	894
Game of Thrones dice cup	stark-dice-cup-1847	893
Overwatch McCree flashbang	mccree-flashbang-from-overwatch-21595	890
B2 bomber glider	b2-stealth-bomber-glider-improved-flight-powered-by-an-elastic-band-13337	886
Star Wars X Wing helmet	x-wing-pilot-helmet-starwars-episode-vii-the-force-awakens-9074	859

The items were uploaded into Cura 15.04.6 (Ultimaker, Geldermalsen, The Netherlands) [29] and the resulting data regarding estimated print time, item weight, and length of the filament used were recorded. In addition, a 3-mm poly lactic acid (PLA) was selected as the filament because it is the most common household consumer 3D printing material and is available from most 3D printing suppliers. PLA has gained prominence, as not only does it demonstrate less warping during printing than other materials such as the second most common 3D printing plastic (ABS, acrylonitrile butadiene styrene), but the emissions during printing are less pungent [30,31]. Furthermore, PLA is made from corn-based resin, making it non-toxic, biodegradable, and able to be produced in environmentally friendly, renewable processes [32]. The items were then categorized into three groups using commercial PLA: (1) those that saved the consumer money when compared to a commercially available alternative product; (2) those that lost money because a less expensive product was available; and (3) those for which there was no alternative product. Items ranged from action figures and masks to non-toy items and cosplay paraphernalia as seen in Table 1.

A commercial price for each product was found primarily on Walmart.com and supplemented using Google Shopping. Associated shipping costs were excluded from the analysis for both purchasing and distributed manufacturing (e.g., no shipping charges included for plastic filament). Following [21], the operating cost (O) for the Lulzbot Mini [33] was calculated using electricity and filament consumption during printing with 15% infill. The average electric rate in the United States was found to be \$0.1267 per kWh for the residential sector [34]. A sensitivity was run on the cost of the filament ($C_{F(source)}$) using:

(1) the most popular filament sold on Amazon.com, $C_{F(filament)}$ is US\$23/kg [35];

(2) the use of a plastic extruder such as a commercial systems (e.g., Filastruder, FilaFab, Noztek, Filabot, EWE, Extrusionbot, Filamaker, Strooder, Felfil, ExtrusionBot) to make filament from commercial PLA pellets (source) $C_{F(pellets)}$ is US\$5.50/kg [36];

(3) the use of a home recyclebot (waste plastic extruder) to convert waste post-consumer waste plastic into filament [37] $C_{F(waste)}$ (which can be as low as a few cents per kg in horizontal recyclebots) for ABS in vertical recyclebot in another study it was found to be US\$2.16/kg [38] and will be used here to be conservative.

The operating cost, O, of a 3D printer was calculated as follows:

$$O = EC_E + \frac{C_F m_f}{1000} [USD], \tag{1}$$

where E is the energy consumed in kWh, C_E is the average rate of electricity in the United States in USD/kWh, C_F is the cost of a given filament in USD/kg, and m_f is the mass of the filament in grams consumed during printing. The marginal savings on each project, Cs, is given by:

$$C_S = C_C - O \qquad [USD], \tag{2}$$

where C_C is the cost of the commercially available product and the marginal percent change, P, between the cost to print a product and the commercially available product was calculated as follows:

$$P = \frac{C_S}{C_C} \times 100[\%], \tag{3}$$

where C_C is the cost for the commercial product at either the high or low price. Finally, the value obtained from a free and open source 3D printable design can be determined from the [39,40] at a specific time (t):

$$V_D(t) = C_S \times P \times N_d(t)[USD], \tag{4}$$

This value is determined by the number of downloads (N_d) during January 2017 and P where is the percent of downloads resulting in a print. It should be pointed out that P is subject to error as downloading a design does not guarantee manufacturing. On the other (more likely) hand, a single download could be fabricated many times, traded via email, memory stick or posted on P2P websites that are beyond conventional tracking. Here, to remain conservative, P is assumed to be 1 and the total savings found over MyMiniFactory in 1 month on the top 100 downloaded items is determined by:

$$V_{DT}(t) = \sum_{i=0}^{100} V_D(t)[USD]. \tag{5}$$

Three case studies are then presented to probe the economics of individual types of toys and games.

2.2. Case Studies

2.2.1. Six Toy Product Comparisons

As is clear from Table 1, the most popular types of 3D printable design are niche-community toys (e.g., gamers, cosplay, etc.). Six more common toy products with existing free designs on MyMiniFactory were selected for more detailed comparisons attractive to a wider audience. Printing costs for these toys were estimated using $24/kg filament and the closest commercial equivalent was found online. The toys are compared visually and the savings percent is calculated by Equation (3), where it is conservatively assumed to be zero (e.g., the electricity cost was ignored).

2.2.2. Lego Case Study

Lego is one of the top five leading toy industry manufacturers with 5% of the market [41]. Lego's signature toy is manufactured to exacting specifications from ABS, a common consumer polymer. In a past study, on average, a single Lego piece costs $0.104 USD, and the average cost of a Lego set without pieces, thereby the cost of the box and printed instructions, is $7.34 [42]. It should be pointed out that, in consideration of their larger size intended for small children, Duplo blocks and thus Duplo sets are more expensive at $0.63 per brick. When grouped into themes, it was found that Lego city-themed and architecture-themed sets had the lowest base cost while Marvel-themed sets had a base cost of $3.61 per piece [42].

The study by Allain is expanded here to look at the cost per block in the 10 most popular Lego kits at Wal-Mart. The cost to print a 3 × 2 block is then determined. The 3D printed blocks are then tested and compared to Lego blocks and generic Lego-compatible commercial blocks on a quality and price basis.

2.2.3. Board Game Comparison

The hobby games market is more than US$1.2 billion, with the hobby game board market having grown 56% from $160 million in 2014 to $250 million in 2015 [43]. ICv2 CEO Milton Griepp has claimed that there is often a transition from digital games to tabletop games when users are more interested in face-to-face interaction [44]. The global board games market is expected to experience a 29% growth over the period from 2017 to 2021, largely due to increased popularity amongst the adult consumer demographic [45]. Three market trends—the growth of organized retail, increase in projects funded through crowdsourcing, and game evolution with time—have been cited as driving forces behind this projected growth. Here again, 3D printing can be used to manufacture board games at home. Technavio points out that 3D printing have been shown to boost the sales of board games [45]. To demonstrate the feasibility and compare the costs to a traditional board game, an open source game, 'Save the planet' board game hosted on Appropedia [46] has been selected. The costs to commercially print out the board and the cards is quantified from a professional print shop and compared to 2D printing on paper by a home printer. Then, the costs for the dice and four player pieces are determined by printing and weighing on a digital scale (±0.1 g) and assuming filament costs from above. These costs are compared to the top ten board games by sales on Amazon [47].

3. Results and Discussion

3.1. Downloaded Value

3.1.1. Microeconomic Advantages of Home Manufacturing

When the cost was calculated for each designed item printed with each of the filaments as described in the Methods section, the recyclebot-produced filament proved to demonstrate the greatest savings for the user. By using an at-home recyclebot or extruder-produced pellets, printing costs can be significantly reduced, and even using more expensive filament found on Amazon demonstrates savings when compared to equivalent, commercially available products. Table 2 shows the cost of each design printed in each of the three filaments as well as the cost of the commercially available product equivalent. On average, printing the items in Amazon filament cost $9.28 while printing in pellets and recyclebot filament cost on average $2.59 and $1.31, respectively. Table 2 shows a clear financial advantage to the consumer in printing items as opposed to purchasing them as shown by the printing cost with recyclebot filament costing on average a mere 3.09% of the cost to purchase. Even when compared to the more expensive Amazon filament, printing costs, on average, only 22% of the cost to purchase. It should also be noted here (and is shown in Section 3.3) that for applications such as toys there is not a noticeable tradeoff between quality and cost of filament. A tuned recyclebot can produce filament with equivalent visual quality to commercial filament.

Table 2. Cost of producing toys and benchmark purchase price (US $).

Design	Mass (g)	Cost: Commercial Filament	Cost: Pellet	Cost: Recyclebot	Commercially Available Alternative (USD)
Pokemon Go aimer	27	$0.66	$0.18	$0.09	$4.99
Clash of Clans barbarian	21	$0.51	$0.14	$0.07	$19.00
Voltron figure	1247	$30.26	$8.44	$4.27	$9.99
Overwatch tracer gun	190	$4.61	$1.29	$0.65	$26.59
Overwatch reaper mask	333	$8.08	$2.25	$1.14	$25.00
Star Wars AT-AT	683	$16.57	$4.62	$2.34	$7.50
Last Word Destiny Hand Cannon	367	$8.91	$2.48	$1.26	$100.00
Overwatch D.VA Light Gun	261	$6.33	$1.77	$0.89	$55.00
Overwatch Reaper Shotgun	269	$6.53	$1.82	$0.92	$40.38
Batman cowl	147	$3.57	$0.99	$0.50	$6.76
Destiny Hawkmoon gun	473	$11.48	$3.20	$1.62	$99.99
Destiny thorn gun	425	$10.31	$2.88	$1.46	$17.99
Star Wars VII Storm Trooper Helmet	948	$23.01	$6.42	$3.25	$20.20
Kylo Ren helmet	550	$13.35	$3.72	$1.88	$4.08
Wall outlet shelf	69	$1.67	$0.47	$0.24	$10.00
Kylo Ren lightsaber	319	$7.74	$2.16	$1.09	$20.00
Blade Runner blaster	249	$6.04	$1.68	$0.85	$12.99
Fallout 3 T45-d helmet	1586	$38.49	$10.73	$5.44	$185.91
Venus de Milo figurine	28	$0.68	$0.19	$0.10	$9.00

Table 2. *Cont.*

Design	Mass (g)	Cost: Commercial Filament	Cost: Pellet	Cost: Recyclebot	Commercially Available Alternative (USD)
Warcraft Frostmourne sword	1233	$29.92	$8.34	$4.23	$133.00
3DR Iris+ quadcopter	255	$6.19	$1.73	$0.87	$13.89
Pieta figurine	164	$3.98	$1.11	$0.56	$16.88
P08 Luger gun	138	$3.35	$0.93	$0.47	$79.00
Game of Thrones iron throne	124	$3.01	$0.84	$0.42	$15.25
The Thinker figurine	92	$2.23	$0.62	$0.32	$22.32
Overwatch McCree Peacemaker gun	301	$7.30	$2.04	$1.03	$69.99
Fallout 4 Pipboy 3000 MkIV	884	$21.45	$5.98	$3.03	$74.99
Articulated lamp	180	$4.37	$1.22	$0.62	$7.97
Overwatch throwing star	13	$0.32	$0.09	$0.04	$6.99
Strong bolt	9	$0.22	$0.06	$0.03	$19.14
Mazinger Z Super Robot	117	$2.84	$0.79	$0.40	$9.99
Sombra pistol	495	$12.01	$3.35	$1.70	$85.00
Han Solo blaster	216	$5.24	$1.46	$0.74	$10.12
Overwatch D.VA headset	65	$1.58	$0.44	$0.22	$18.99
Overwatch Mercy staff	1768	$42.90	$11.96	$6.06	$165.89
Overwatch Mercy blaster	411	$9.97	$2.78	$1.41	$19.49
Gears of War Chainsaw gun	1556	$37.76	$10.53	$5.33	$199.99
Destiny Duke MK.44 gun	325	$7.89	$2.20	$1.11	$125.00
Fallout 4 Laser pistol	1322	$32.08	$8.95	$4.53	$89.99
Overwatch loot box	124	$3.01	$0.84	$0.42	$8.98
Cat at British Museum	62	$1.50	$0.42	$0.21	$75.99
Halo 5 assault rifle	2295	$55.69	$15.53	$7.86	$50.00
Portal gun	1528	$37.08	$10.34	$5.24	$279.99
Clash of Clans figurine	2	$0.05	$0.01	$0.01	$29.99
Harry Potter elder wand	17	$0.41	$0.12	$0.06	$4.79
Groot flower pot	57	$1.38	$0.39	$0.20	$8.99
Melted Darth Vader mask	172	$4.17	$1.16	$0.59	$8.00
Fallout 4 10mm pistol	793	$19.24	$5.37	$2.72	$24.99
Starcraft Kerrigan statue	112	$2.72	$0.76	$0.38	$55.79
Destiny ghost	116	$2.81	$0.78	$0.40	$7.00
Lich king figurine	487	$11.82	$3.30	$1.67	$15.00
Ant Man helmet	367	$8.91	$2.48	$1.26	$44.59
Game of Thrones House emblem	34	$0.83	$0.23	$0.12	$9.95
Clash of Clans king figure	24	$0.58	$0.16	$0.08	$19.00
Tooth toothbrush holder	120	$2.91	$0.81	$0.41	$2.00
Halo 3 ODST helmet	1118	$27.13	$7.57	$3.83	$9.99
Fallout 4 protectron figure	98	$2.38	$0.66	$0.34	$24.99
Discobolus figurine	35	$0.85	$0.24	$0.12	$7.25
Anonymous mask	383	$9.29	$2.59	$1.31	$4.95
Guardian of the Galaxy Star Lord mask	351	$8.52	$2.38	$1.20	$5.99
Pokeball	221	$5.36	$1.50	$0.76	$1.00
Witcher 3 wall plaque	100	$2.43	$0.68	$0.34	$45.00
Overwatch widowmaker rifle	1364	$33.10	$9.23	$4.67	$145.89
Fallout 4 combat rifle and shotgun	351	$8.52	$2.38	$1.20	$85.00
Michelangelo's David	75	$1.82	$0.51	$0.26	$29.93
Triceratops skull	66	$1.60	$0.45	$0.23	$40.00
Skull ring	5	$0.12	$0.03	$0.02	$9.95
Star Wars NN-44 Rey's Blaster	141	$3.42	$0.95	$0.48	$45.00
Joker mask	751	$18.22	$5.08	$2.57	$11.56
Fork/spoon support for disability	17	$0.41	$0.12	$0.06	$11.00
Gryffindor coat of arms	48	$1.16	$0.32	$0.16	$8.00
Pokemon figurines	63	$1.53	$0.43	$0.22	$30.00
Rubik's cube	12	$0.29	$0.08	$0.04	$4.99
Fallout 4 Kellogg's pistol	152	$3.69	$1.03	$0.52	$100.00
Dobby the Elf figurine	365	$8.86	$2.47	$1.25	$31.36
Statue of Liberty figurine	85	$2.06	$0.58	$0.29	$6.44
Nutcracker	61	$1.48	$0.41	$0.21	$4.99
Minions figurines	121	$2.94	$0.82	$0.41	$26.80
Destiny sleeper simulant	1274	$30.92	$8.62	$4.37	$211.66
Frozen Elsa figurine	33	$0.80	$0.22	$0.11	$7.99
Star Wars storm trooper rifle	486	$11.79	$3.29	$1.67	$170.00
Vitruvian Man scuplture	12	$0.29	$0.08	$0.04	$31.49
Robocop ED209 figure	736	$17.86	$4.98	$2.52	$21.99
Harry Potter wand	8	$0.19	$0.05	$0.03	$13.55
Cable guards	7	$0.17	$0.05	$0.02	$1.25
Game of Thrones dice cup	59	$1.43	$0.40	$0.20	$8.32
B2 bomber glider	20	$0.49	$0.14	$0.07	$10.57
Star Wars X Wing helmet	949	$23.03	$6.42	$3.25	$41.40
Hawkmoon #2	621	$15.07	$4.20	$2.13	$99.99
Destiny ghost #2	48	$1.16	$0.32	$0.16	$7.00
Average	382.29	$9.28	$2.59	$1.31	$42.44
Total	34,406	$834.93	$232.83	$117.91	$3819.59

Table 3 shows the cost and percent savings calculated for each design in each of the three filament types when compared to their respective commercially available alternative products. As expected based on the results in Table 2, printed items using recyclebot-made filament saved the user significantly more money than printed items using commercial or pellet filament; however, each of the three filament types on average saved the user over 50% of the cost of commercially available alternative toys. When items that cost more to print than purchase were removed, the average cost and percent savings were nearly 75% when using commercial filament and over 90% for pellet-extruded filament and recyclebot filament.

Table 3. Cost savings for individual toys (US$) and percent (%) savings.

Design	Cost Savings (USD)			Percent Savings (%)		
	Commercial Filament	Pellets	Recyclebot	Commercial Filament	Pellets	Recyclebot
Pokemon Go aimer	$4.33	$4.81	$4.90	87%	96%	98%
Clash of Clans barbarian	$18.49	$18.86	$18.93	97%	99%	100%
Voltron figure	$(20.27)	$1.55	$5.72	−203%	16%	57%
Overwatch tracer gun	$21.98	$25.30	$25.94	83%	95%	98%
Overwatch reaper mask	$16.92	$22.75	$23.86	68%	91%	95%
Star Wars AT-AT	$(9.07)	$2.88	$5.16	−121%	38%	69%
Last Word Destiny Hand Cannon	$91.09	$97.52	$98.74	91%	98%	99%
Overwatch D.VA Light Gun	$48.67	$53.23	$54.11	88%	97%	98%
Overwatch Reaper Shotgun	$33.85	$38.56	$39.46	84%	95%	98%
Batman cowl	$3.19	$5.77	$6.26	47%	85%	93%
Destiny Hawkmoon gun	$88.51	$96.79	$98.37	89%	97%	98%
Destiny thorn gun	$7.68	$15.11	$16.53	43%	84%	92%
Star Wars VII Storm Trooper Helmet	$(2.81)	$13.78	$16.95	−14%	68%	84%
Kylo Ren helmet	$(9.27)	$0.36	$2.20	−227%	9%	54%
Wall outlet shelf	$8.33	$9.53	$9.76	83%	95%	98%
Kylo Ren lightsaber	$12.26	$17.84	$18.91	61%	89%	95%
Blade Runner blaster	$6.95	$11.31	$12.14	53%	87%	93%
Fallout 3 T45-d helmet	$147.42	$175.18	$180.47	79%	94%	97%
Venus de Milo figurine	$8.32	$8.81	$8.90	92%	98%	99%
Warcraft Frostmourne sword	$103.08	$124.66	$128.77	78%	94%	97%
3DR Iris+ quadcopter	$7.70	$12.16	$13.02	55%	88%	94%
Pieta figurine	$12.90	$15.77	$16.32	76%	93%	97%
P08 Luger gun	$75.65	$78.07	$78.53	96%	99%	99%
Game of Thrones iron throne	$12.24	$14.41	$14.83	80%	94%	97%
The Thinker figurine	$20.09	$21.70	$22.00	90%	97%	99%
Overwatch McCree Peacemaker gun	$62.69	$67.95	$68.96	90%	97%	99%
Fallout 4 Pipboy 3000 MkIV	$53.54	$69.01	$71.96	71%	92%	96%
Articulated lamp	$3.60	$6.75	$7.35	45%	85%	92%
Overwatch throwing star	$6.67	$6.90	$6.95	95%	99%	99%
Strong bolt	$18.92	$19.08	$19.11	99%	100%	100%
Mazigner Z Super Robot	$7.15	$9.20	$9.59	72%	92%	96%
Sombra pistol	$72.99	$81.65	$83.30	86%	96%	98%
Han Solo blaster	$4.88	$8.66	$9.38	48%	86%	93%
Overwatch D.VA headset	$17.41	$18.55	$18.77	92%	98%	99%
Overwatch Mercy staff	$122.99	$153.93	$159.83	74%	93%	96%
Overwatch Mercy blaster	$9.52	$16.71	$18.08	49%	86%	93%
Gears of War Chainsaw gun	$162.23	$189.64	$194.66	81%	95%	97%
Destiny Duke MK.44 gun	$117.11	$122.80	$123.89	94%	98%	99%
Fallout 4 Laser pistol	$57.91	$81.04	$85.46	64%	90%	95%
Overwatch loot box	$5.97	$8.14	$8.56	66%	91%	95%
Cat at British Museum	$74.49	$75.57	$75.78	98%	99%	100%
Halo 5 assault rifle	$(5.69)	$34.47	$42.14	−11%	69%	84%
Portal gun	$242.91	$269.65	$274.75	87%	96%	98%
Clash of Clans figurine	$29.94	$29.98	$29.98	100%	100%	100%
Harry Potter elder wand	$4.38	$4.67	$4.73	91%	98%	99%
Groot flower pot	$7.61	$8.60	$8.79	85%	96%	98%
Melted Darth Vader mask	$3.83	$6.84	$7.41	48%	85%	93%
Fallout 4 10mm pistol	$5.75	$19.62	$22.27	23%	79%	89%
Starcraft Kerrigan statue	$53.07	$55.03	$55.41	95%	99%	99%
Destiny ghost	$4.19	$6.22	$6.60	60%	89%	94%
Lich king figurine	$3.18	$11.70	$13.33	21%	78%	89%
Ant Man helmet	$35.68	$42.11	$43.33	80%	94%	97%
Game of Thrones House emblem	$9.12	$9.72	$9.83	92%	98%	99%

Table 3. *Cont.*

Design	Cost Savings (USD)			Percent Savings (%)		
	Commercial Filament	Pellets	Recyclebot	Commercial Filament	Pellets	Recyclebot
Clash of Clans king figure	$18.42	$18.84	$18.92	97%	99%	100%
Tooth toothbrush holder	$(0.91)	$1.19	$1.59	−46%	59%	79%
Halo 3 ODST helmet	$(17.14)	$2.42	$6.16	−172%	24%	62%
Fallout 4 protectron figure	$22.61	$24.33	$24.65	90%	97%	99%
Discobolus figurine	$6.40	$7.01	$7.13	88%	97%	98%
Anonymous mask	$(4.34)	$2.36	$3.64	−88%	48%	73%
Guardian of the Galaxy Star Lord mask	$(2.53)	$3.61	$4.79	−42%	60%	80%
Pokeball	$(4.36)	$(0.50)	$0.24	−436%	−50%	24%
Witcher 3 wall plaque	$42.57	$44.32	$44.66	95%	98%	99%
Overwatch widowmaker rifle	$112.79	$136.66	$141.22	77%	94%	97%
Fallout 4 combat rifle and shotgun	$76.48	$82.62	$83.80	90%	97%	99%
Michelangelo's David	$28.11	$29.42	$29.67	94%	98%	99%
Triceratops skull	$38.40	$39.55	$39.77	96%	99%	99%
Skull ring	$9.83	$9.92	$9.93	99%	100%	100%
Star Wars NN-44 Rey's Blaster	$41.58	$44.05	$44.52	92%	98%	99%
Joker mask	$(6.66)	$6.48	$8.99	−58%	56%	78%
Fork/spoon support for disability	$10.59	$10.88	$10.94	96%	99%	99%
Gryffindor coat of arms	$6.84	$7.68	$7.84	85%	96%	98%
Pokemon figurines	$28.47	$29.57	$29.78	95%	99%	99%
Rubik's cube	$4.70	$4.91	$4.95	94%	98%	99%
Fallout 4 Kellogg's pistol	$96.31	$98.97	$99.48	96%	99%	99%
Dobby the Elf figurine	$22.50	$28.89	$30.11	72%	92%	96%
Statue of Liberty figurine	$4.38	$5.86	$6.15	68%	91%	95%
Nutcracker	$3.51	$4.58	$4.78	70%	92%	96%
Minions figurines	$23.86	$25.98	$26.39	89%	97%	98%
Destiny sleeper simulant	$180.74	$203.04	$207.29	85%	96%	98%
Frozen Elsa figurine	$7.19	$7.77	$7.88	90%	97%	99%
Star Wars storm trooper rifle	$158.21	$166.71	$168.33	93%	98%	99%
Vitruvian Man sculpture	$31.20	$31.41	$31.45	99%	100%	100%
Robocop ED209 figure	$4.13	$17.01	$19.47	19%	77%	89%
Harry Potter wand	$13.36	$13.50	$13.52	99%	100%	100%
Cable guards	$1.08	$1.20	$1.23	86%	96%	98%
Game of Thrones dice cup	$6.89	$7.92	$8.12	83%	95%	98%
B2 bomber glider	$10.08	$10.43	$10.50	95%	99%	99%
Star Wars X Wing helmet	$18.37	$34.98	$38.15	44%	84%	92%
Hawkmoon #2	$84.92	$95.79	$97.86	85%	96%	98%
Destiny ghost #2	$5.84	$6.68	$6.84	83%	95%	98%

Upon review, the items that cost more to print than to purchase (e.g., potentially lost the user money) when compared to commercially available and comparable products were often of noticeably higher value. For example, in Figure 1, the Game of Thrones inspired dice cup shown rendered in Cura is not commercially available but was compared to simple dice cups.

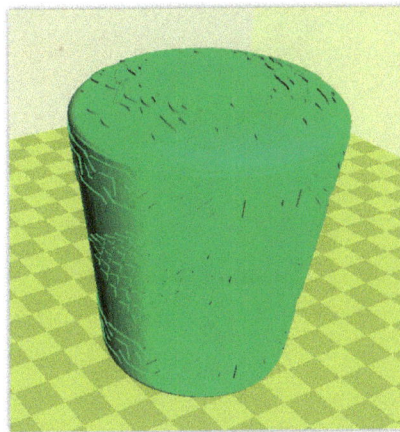

Figure 1. Cura rendering of details of freely available design of a Game of Thrones dice cup.

The majority of the items that cost more to print than to purchase were specialized cosplay items. The term "cosplay" was added to the Oxford English Dictionary in 2008 as "the practice of dressing up as a character from a film, book, or video game, especially one from the Japanese genres of manga or anime" [48]. The appeal of cosplay is largely the expression of individualism within a shared community [49]. Not only does at-home additive manufacturing grant individual users the power to design and build completely customized items, but repositories such as MyMiniFactory provide an outlet for users to share designs and inspire one another in their creativity. Items in this study that did not save the user money when printed in all filaments such as the Voltron figure had dozens of specially designed parts that together made up a customized product of arguably significantly higher quality than the simple commercial alternative.

In addition, 6% of designs had no commercially available alternative either on Walmart.com nor offered by individual makers. Many of these were highly detailed cosplay items that came from specific games. While a cost saving could not be calculated, these items only highlight 3D printing's ability to allow users unlimited creativity in their work, and MyMiniFactory's data for these designs' downloads prove that these items are highly desired by the 3D printing and cosplaying community. Figure 2 shows the average percent savings of the cumulative designs with commercially available alternative elements based on the filaments used to print the items. Items printed with recyclebot-made filament demonstrated the greatest percent savings at 93% when all items were considered and 97% when only items that saved the user money were considered. It is clear that using recycled waste plastic would save consumers more than 90% of the costs of the conventionally-manufactured cost regardless of the circumstances. As 3D printing continues to become a more widely spread at-home activity and investment, filament technology may progress such that many users will utilize waste plastic as opposed to commercially available filament. Given the multi-billion status that is the toy industry, consumers may begin printing toys, games and specialty items in mass, saving substantial amounts of money in comparison to commercially available products.

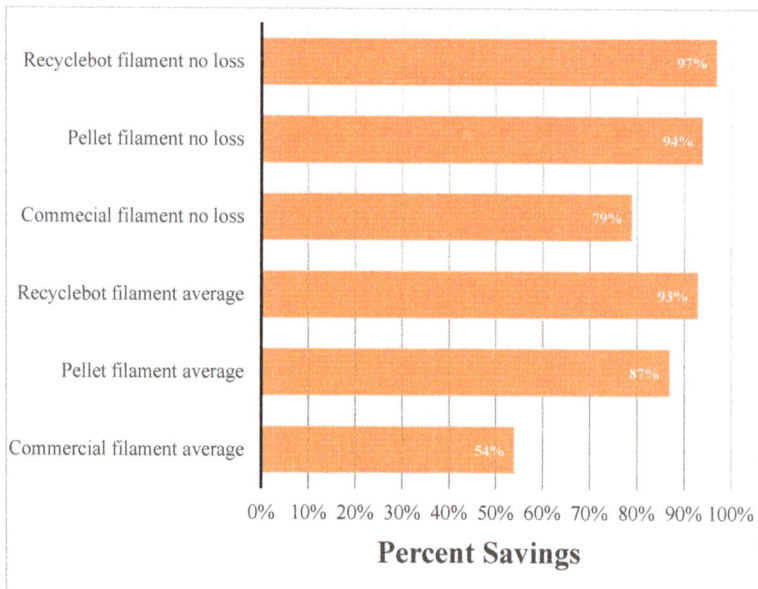

Figure 2. Bar chart showing the percent savings of printing designs with various filament types.

3.1.2. Macroeconomic Impact of Home Manufacturing

The total results from Table 3 are the most striking. Over a single month period, MyMiniFactory saved its users more than $5 million in avoided consumer purchases for only the top 100 downloaded designs. This assumes filament costs $20/kg or less, which is a reasonable assumption as the most popular filament is $23/kg and there are many commercial suppliers (even available on Amazon) that are selling filament for less than $20/kg as well as the much lower costs observed with pellet extruded filament or recyclebot manufactured filament. Extrapolated over a year, these results indicate that MyMiniFactory, just a single 3D printing repository among dozens, is saving consumers over $60 million a year in offset purchases. Again, this value is conservative as many of the designs discarded from the analysis here would be more (not less) expensive to acquire commercially. In addition, this only considers 100 designs. MyMiniFactory currently has 26,355 published designs as of 30 April 2017 so this study only looked at 0.38%. In addition, this represents, but a small fraction of the overall freely available designs, which is at least several million [21].

It can be safely concluded that the open source 3D printing community is already having more than $100 million/year on the toy and game market. As the microeconomic savings are significant for individual consumers and the number of desktop 3D printer users continues to climb, this impact can be expected to grow.

3.2. Six Example Printable Toys

Although the toys can be made for less than purchased commercially, they may not be of the same value. Figure 3 shows the visual results of six common household toys to probe this effect. The free design 3D printable version is shown on the left and the commercial version in shown on the right. First, the mini travel chess set available from the largest online retailer, Amazon, results in a 90% savings if 3D printed. The color difference actually provides a visual advantage to 3D printing. Although it should be pointed out that identical (or different) color schemes could be enabled simply by using different colors of plastic filament for the different components of a toy. For those that find the wood more aesthetically appealing, there is already a wide range of wood filaments on the market and other biocomposites [50], and recent work indicates that even wood waste can be converted to 3D printer filament [51]. Home manufacturing of toys is also more economic when compared to a dedicated toy seller like ToysRUs. The shape puzzle available from ToysRUs results in a 88% savings and the toy truck 79% savings. In both cases, the commercial versions have a higher degree of color variance than shown, but this could be overcome by using more colors of filament and/or changing filaments during a single print. For more complex toys, like the action figures, 3D printing still results in a savings, although lower (e.g., 41%) values. In addition to mass manufactured toys, wood toys made available on sites like Etsy can also be replaced with distributed manufacturing using 3D printing with substantial cost savings. For example, the wood puzzle star available on Etsy can be 3D printed for 82% savings as can the math spinner toy for 90% savings. It should be noted here that the relative cost savings can be heavily influenced not only by the filament selection but also the infill percentage. For example, when comparing 3D printed toys to wood based objects, the environmental impact can be lower when low infill settings are used, and, in general for non-solid plastic products, 3D printed ones have a lower impact than conventionally manufactured ones [52,53].

Figure 3. Visual comparison of open source 3D printed toys and their commercial mass manufactured equivalents. Costs and percent savings are shown for each toy.

Making a more careful comparison of some of the example toys shown in Figure 3, it is clear that these toys both appear and are actually different than the commercial counterparts (e.g., design, shape, and color) although the functionality (e.g., play) is similar or identical. The difference is perhaps most stark with the action figure (e.g., Captain America vs. BloodShot). There is currently no equivalent quality BloodShot action figure and the intellectual property surrounding the example is owned by Valliant Comics. The play with either action figure is identical (or nearly identical as super powers for both comic book characters arose from former soldiers being injected with super solider formula and nanites, respectively). However, a specific branded toy has a value to the consumer. This value can be acquired by the consumer by customizing the toy to fit their need (e.g., add minor changes to the existing action figure and/or repaint it) as well as print accessories such as a Captain America shield, which is already available on MyMiniFactory. The intellectual property concerns of home manufacturers doing such modifications are left for future work.

3.3. Lego Analysis

Lego is well known in the open source maker community (e.g., commons based peer production using a Lego-built 3D printing/Milling machine [54]). There are already hundreds of Lego designs available and customized OpenSCAD code generators for various Lego-compatible blocks. As can be seen in Figure 4, it is possible to fabricate Lego compatible bricks for less money than purchasing them with any type of ABS filament. It should be noted that the fit of the Lego blocks are superior to the generic compatible block and all of the 3D printed blocks on their first run. The fit of the 3D printed blocks, however, can be adjusted by the individual consumer to make blocks easier to disassemble (e.g., for weaker hands) or tighter (e.g., to make more permanent structures). The quality of the 3D printed parts can also be a key determinant in demand. Care must be taken by home toy manufacturers to ensure both the polymer used as well as the infill is appropriate for the toy being fabricated. In addition, there are some inherent limitations on the visual quality of FFF 3D printed parts. As can be seen in Figure 4, for the three unsmoothed 3D printed blocks lines can be observed on the z-axis. However, if this is important to the home user, ABS can be smoothed (bottom right block in Figure 4) with acetone, rendering a block very close to the visual quality of the generic block and removing print lines. In addition, 3D printed blocks can be made that are not available from Lego as shown as the example of the Lego to Lincoln Logs adapter blocks rendered by Cura [55] in Figure 5. It should be noted that several runs may be needed by the home manufacturer of Lego compatible blocks to obtain an ideal fit for the users, which would contributed to higher costs, although to a small degree. For example, it might take five tries to get the perfect fit, but then the settings could be used to print out a standard set of 1000 blocks so the trials needed would represent a minor loss and cost (e.g., 0.5%). For many other toys, such careful tolerances are not necessary.

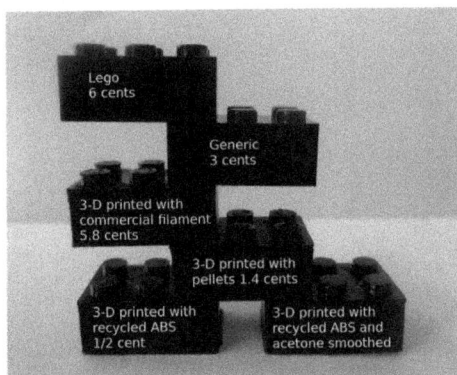

Figure 4. Photograph of 2 × 3 toy bricks: Lego brick, Lego-compatible generic brick, 3D printed commercial ABS and 3D printed recycled acrylonitrile butadiene styrene (ABS) bothnatural and acetone smoothed with costs.

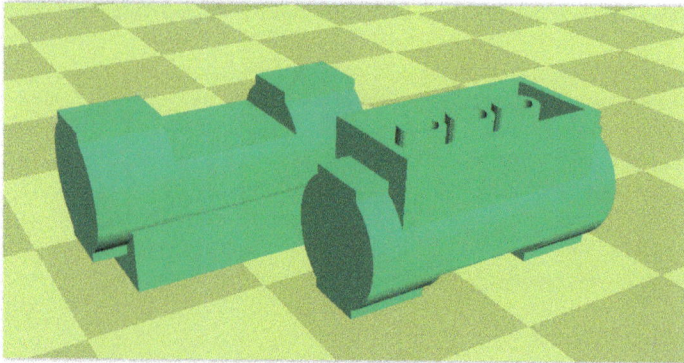

Figure 5. Adapterz LLC free design of a Lego to Lincoln Log adapter [55] shown in Cura.

2 × 3 3D printed Lego compatible blocks have a mass of 2.5 g. To provide a fair comparison to a commercially available Lego-compatible generic brick set [56], the costs of fabricating 1000 2 × 3 Lego compatible bricks is shown with the various sources shown in Figure 6. In reviewing the top 10 most popular Lego sets on Walmart.com, it was found that the average cost per piece was $0.075. The average cost for these kits on Amazon.com was $74.82 with simple Lego blocks costing $60.00. Generic building blocks cost less at $29.99 [56]. When estimating the cost using 3D printed blocks, a comparable kit printed with commercial filament would cost $57.50. A kit printed from pellets would cost $13.75 and one printed from recyclebot-made filament would cost only $5.40. In consideration of *Wired*'s cost estimate of a single Lego brick [42], a 1000-piece kit would cost $104, nearly double that of the kit printed with the more expensive commercial filament. Interestingly, with the cost of RepRap 3D printer kits now breaking the $100 cost barrier (the Startt 3D printer from iMakr (London, England) is currently selling for $99.99) [57], only roughly two sets of Legos need to be printed with pellet-made filament to recover the cost of the 3D printer. However, it should be noted that such kits still require substantial technical competence from the consumer.

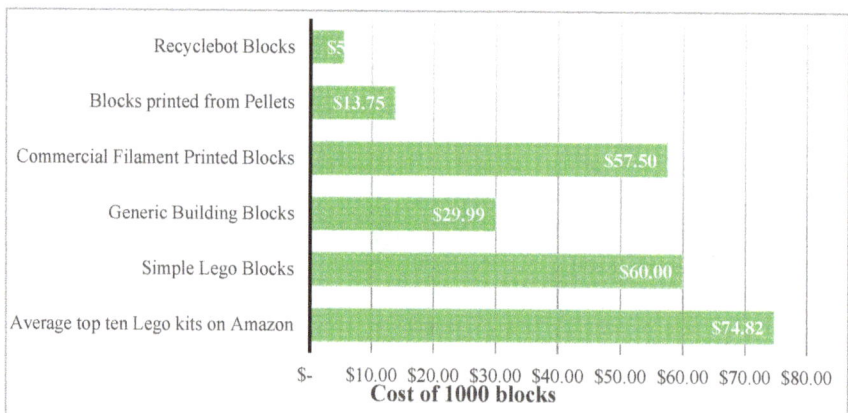

Figure 6. The costs of fabricating 1000 2 × 3 Lego compatible bricks or purchasing.

3.4. Board Game

Save the Planet Board Game is a free and open-source DIY cooperative board game [46] shown in Figure 7. In the game, players work together to save the planet to win the game while learning how to save the planet in real life. Users must 2D print out the game board and card decks for the beginner and advanced options, respectively. This can be done on 8 × 11 inch office paper (4 × 11 cents/page is 44 cents) and laminated to be glued to cereal boxes or printed directly on cardstock (1 × 69 cents) so the total 2D printing cost is $1.13. The beginner option has game pieces for simple acts of environmental benefit and is appropriate for children 4 and up and the advanced option is for teens and adults interested in more scalable impact. This game is open-source so like the RepRap, users are encouraged to build on it—make it better, add more "good deeds", make a local deed list and make more advanced derivatives.

Figure 7. Open-source fully 2D and 3D printable do-it-yourself (DIY) cooperative *Save the Planet Board Game* in mid-play.

The following needs to be printed (summarized in Table 4): a card holder, dice and four mini-figures. The mini-figurines are completely up to the interests of the user and, with thousands of designs to choose from, enable some creativity of the user. There is also some flexibility in the size of the figures that will affect the cost and found items (e.g., a small stone) could reduce the cost of the figures to zero but is of lower value. Here, in order to demonstrate the range of options a fox, The Thinker at the Musée Rodin France, a figurine from another open board game, and a superman bust are used.

Table 4. 3D and 2D printed components, source and cost of *Save the Planet Board Game* assuming commercial filament and cardstock.

Item	Source (myminifactory.com/object/)	Mass/Sheets	Cost [USD]
Card holder	card-holder-for-save-the-planet-board-game-35620	39.15	0.90
Dice	the-magnificent-dice-27043	1.68	0.04
The Thinker	the-thinker-at-the-muse-rodin-france-2127	13.86	0.32
Fox	fox-support-free-5865	8.04	0.18
Open board game figurine	open-board-game-figurine-6013	7.50	0.17
Superman bust	superman-bust-3518	6.46	0.15
Game Boards	2D printing costs	5 sheets	1.13
	Total		2.89

The top 10 best-selling board games on Amazon as of 3 May 2017 averaged in price at $18.18 with a range of $23.22, where the most expensive game was $31.99 and the least expensive was $8.77 [47]. The 3D and 2D printed components, source and cost of *Save the Planet Board Game* costs less than $3 as shown in Table 4, representing a 67%–91% savings. However, in this case, it is not an apple to apple comparison for identical games, which would come with a risk of intellectual property infringement. Instead, this analysis was for comparing a custom board game (with potentially greater value) to a generic board game. The closest analog of the example game is *A Beautiful Place* that costs $16.49 on Amazon and is significantly less sophisticated, although similar in game play.

Already, many custom games are available in the open source community that have identical game play to conventionally manufactured and sold games, but are superior in some way. For example, many conventionally-manufactured *BattleShip* games are available on the market in the general price range of the most popular board games. All of them have relatively simple injection molded ships. A 3D printable game has been developed (battlefleet-star-wars-vs-star-trek-33840), which has identical game play but uses fan art mockups of *Star Wars* and *Star Trek* ships. The printed game also comes in 3D printable carrying cases for easy mobile play and is potentially of greater value than the Battleship games on the market for *Star Wars/Star Trek* fans.

3.5. Discussion

As the use of 3D printing has shifted from rapid prototyping in industry to production [58], research has shown DIY in-home manufacturing could easily justify the capital costs of a 3D printer with consumer items [21], and this study shows that even when a home printer is focused on making only games and toys, it is clear consumers can generate higher value items for less money than is currently available commercially. It should be pointed out here that these conclusions are in general conservative as this study focused only on relatively simple toys and games that required minimal assembly. Coupling open-source electronics to 3D printed toys (e.g., Arduino-driven 3D printable quad copters and VR headsets) enables far more sophisticated toys with higher values to be produced in the home. One area where this study was not conservative was in the estimation of failed print waste. Here, it was assumed to be zero as in the vast majority of cases printing a known 3D printable part (e.g., guaranteed 3D printable by MyMiniFactory) on an auto-calibrating/bed-leveling 3D printer (e.g., Lulzbot Taz or Mini) nearly always results in a successful print. This is not always the case, as a previous study using a home-built RepRap 3D printer estimated a 20% failure rate [20]. It should be noted, however, that this is only the case on the first prints from an inexperienced user on a much less sophisticated RepRap than are currently available to build, purchase assembled or buy in kit form. Future work could provide a more robust estimate of print failure rates by studying a large group of home 3D printer users.

3D printing gives consumers the unique ability to relatively easily fabricate products only for themselves, which may disrupt manufacturing in a wide array of markets [59–61]. Rifkin argues distributed manufacturing with 3D printers can lead to a zero marginal cost society [62]. It is unclear if the inconvenience of 3D printing yourself will not overcome the convenience of producing exceptionally low cost but high value bespoke products in one's own home; however, it is clear that it will have an impact on global value chains [63,64]. There have been a number of studies concluding that 3D printing will continue to have an increased impact on society [65,66] both in the developed and developing world [67,68]. It is clear that 3D printing will play a major role in the rapidly emerging business models based on open source hardware [69] and open innovation [70]. In addition, even the humble toys here can have added value in the context of medicine when used as therapeutic aides (e.g., magnetic resonance imaging (MRI) visual demonstration tool [71]). Future open-source toy designs can also begin to take advantage of printing [72] or home milling electronic components [73], multimaterials [73,74] and 4D printing smart materials [75] to increase the toy complexity and playability. As the complexity increases, methods will be needed to ensure that such home manufactured toys meet known standards. Work in this area has already commenced [76].

In addition, there has been significant efforts targeting quality control for low-cost FFF 3D printers both informally [77–79] and formally [80–82] to enable real-time control, which when widespread will only enhance the conclusions in this study.

Future work is needed in both the technical areas to continue to reduce costs and improve reliability of open source 3D printers as well as to expand the economic analysis presented in this study. Future work can probe the time needed to design toys using open source software to determine ROIs for designers. In addition, more granular values can be obtained by doing sensitivity analysis on 3D printer user time for setting up a print and on routine maintenance (as well as associated costs). Future work is also needed to quantify the value to the prosumer of making his or her own toys. In this study, the value of the assembly time was considered to be zero as most of the toys were relatively simple to assemble. However, for more complex toys, there would be a time investment. Normal manufacturing considers this assembly time a cost, while, in this case, the assembly may be treated as part of the value of the toy itself in the same way that assembling a Lego kit is part of the game play. Finally, further work is needed to address the quality of recycling plastic for filament. This study showed that the largest savings could be found for distributed home manufacturing for products using recycled waste plastic filament. However, each time a thermoplastic is recycled, the mechanical properties are degraded. Cruz et al. have begun to investigate this phenomena's impact on open source 3D printing [83], but considerably more work is needed in the area over the complete array of polymers used in FFF.

With both the continued decrease in the cost of open source 3D printers and 3D printing filament along with the increase in the number and quality of free designs, it appears clear that consumer DIY manufacturing in the home [84] is set to have a significant impact on the toy and game markets in the future.

4. Conclusions

This study quantified the savings for consumers that utilize free and open source designs with a desktop fused filament 3D printer to fabricate their own toys and games. The cost of the filament was the largest variable controlling savings per product; however, each of the three filament types analyzed here on average saved the user over 75% of the cost of commercially available true alternative toys and over 90% for pellet-extruded filament and recyclebot-made filament. The reduced visual quality (e.g., 3D printer lines) was offset when compared to commercially available and comparable products because the 3D printable version could contain customization and increased complexity that created noticeably higher value for consumers. Over a single month period, MyMiniFactory saved its users more than $5 million in avoided consumer purchases for only the top 100 downloaded designs. These results indicate that MyMiniFactory, just a single 3D printing repository among dozens, is saving consumers well over $60 million a year in offset purchases. The specific case studies found that most common toy savings fell: 40%–90% in cost savings when using the most expensive filament. These cost savings came with ability to make never before seen toys. For example, although the cost of Lego blocks could be cut from 6 cents/block to about 0.5 cents per block using recycled filament, the real strength of 3D printing blocks is to make exactly what the consumer wants Lego compatible. Professional looking games fostering more creativity, customization and in-depth thought of the consumer can also be manufactured at home for a small fraction of purchasing them directly. The results of this study make it clear that consumers can generate higher value items for less money using an open source distributed manufacturing paradigm. With both the continued decrease in the cost of open source 3D printers and 3D printing filament along with the increase in the number and quality of free and open source designs, it appears clear that consumer DIY manufacturing is set to have a significant impact on the toy and game markets in the future.

Acknowledgments: The authors would like to thank MyMiniFactory for supplying data.

Author Contributions: Joshua M. Pearce conceived and designed the study; Romain W. Kidd and Emily E. Petersen performed the data collection; Romain W. Kidd contributed data and Joshua M. Pearce contributed analysis tools; and all authors analyzed the data and wrote the paper. All authors have read and approved the final manuscript.

Conflicts of Interest: The authors declare no conflict of interest.

Acronyms and Symbols

ABS	acrylonitrile butadiene styrene
AM	additive manufacturing
C_C	cost of the commercially available product USD
C_E	average rate of electricity in the United States in USD/kWh
C_F	cost of a given filament in USD/kg
Cs	marginal savings on each project USD
E	energy consumed in kWh
FFF	fused filament fabrication
m_f	mass of the filament in grams
N_d	number of downloads
O_L	operating cost
P	marginal percent change between the cost to print a product and the commercially available product
PLA	poly lactic acid
RepRap	self-REPlicating RAPid prototyper 3D printer
$V_D(t)$	downloaded substitution valuation USD

1. Sells, E.; Bailard, S.; Smith, Z.; Bowyer, A.; Olliver, V. RepRap: The Replicating Rapid Prototyper-Maximizing Customizability by Breeding the Means of Production. In Proceedings of the World Conference on Mass Customization and Personalization, Cambridge, MA, USA, 7–10 October 2007.

2. Jones, R.; Haufe, P.; Sells, E.; Iravani, P.; Olliver, V.; Palmer, C.; Bowyer, A. RepRap-the Replicating Rapid Prototyper. *Robotica* **2011**, *29*, 177–191. [CrossRef]

3. Bowyer, A. 3D Printing and Humanity's First Imperfect Replicator. *3D Print. Addit. Manuf.* **2014**, *1*, 4–5. [CrossRef]

4. Gibb, A.; Abadie, S. *Building Open Source Hardware: DIY Manufacturing for Hackers and Makers*, 1st ed.; Addison-Wesley Professional: Boston, MA, USA, 2014.

5. Rundle, G. *A Revolution in the Making*; Simon and Schuster: New York, NY, USA, 2014.

6. Wohlers, T. *Wohlers Report 2016: 3D Printing and Additive Manufacturing State of the Industry Annual Worldwide Progress Report*; Wohlers Associates Inc.: Fort Collins, CO, USA, 2016.

7. Frauenfelder, M. *Make: Ultimate Guide to 3D Printing 2014: Maker Media*; O'Reilly Inc.: Sebastopol, CA, USA, 2013.

8. Moilanen, J.; Vaden, T. 3D Printing Community and Emerging Practices of Peer Production. *First Monday* **2013**. [CrossRef]

9. Pearce, J.M. Building Research Equipment with Free, Open-Source Hardware. *Science* **2012**, *337*, 1303–1304. [CrossRef] [PubMed]

10. Pearce, J. *Open-Source Lab: How to Build Your Own Hardware and Reduce Research Costs*, 1st ed.; Elsevier: Waltham, MA, USA, 2014.

11. Baden, T.; Chagas, A.; Marzullo, T.; Prieto-Godino, L.; Euler, T. Open Laware: 3-D Printing Your Own Lab Equipment. *PLoS Biol.* **2015**, *13*, e1002086.

12. Coakley, M.; Hurt, D.E. 3D Printing in the Laboratory Maximize Time and Funds with Customized and Open-Source Labware. *J. Lab. Autom.* **2016**, *21*, 489–495. [CrossRef] [PubMed]

13. Kentzer, J.; Koch, B.; Thiim, M.; Jones, R.W.; Villumsen, E.; May. An Open Source Hardware-Based Mechatronics Project: The Replicating Rapid 3-D Printer. In Proceedings of the 2011 4th International Conference on Mechatronics, Kuala Lumpur, Malaysia, 17–19 May 2011; pp. 1–8.

14. Irwin, J.L.; Oppliger, D.E.; Pearce, J.M.; Anzalone, G. Evaluation of RepRap 3D Printer Workshops in K-12 STEM. In Proceedings of the 122nd ASEE Annual Conference and Exposition, Seattle, MA, USA, 14–17 June 2015.

15. Gonzalez-Gomez, J.; Valero-Gomez, A.; Prieto-Moreno, A.; Abderrahim, M. A new open source 3d-printable mobile robotic platform for education. In *Advances in Autonomous Mini Robots*; Springer: Berlin/Heidelberg, Germany, 2012; pp. 49–62.

16. Grujović, N.; Radović, M.; Kanjevac, V.; Borota, J.; Grujović, G.; Divac, D. 3D Printing Technology in Education Environment. In Proceedings of the 34th International Conference on Production Engineering, Nis, Serbia, 28–30 September 2011; pp. 29–30.

17. Schelly, C.; Anzalone, G.; Wijnen, B.; Pearce, J.M. Open-source 3-D printing technologies for education: Bringing additive manufacturing to the classroom. *J. Visual Lang. Comput.* **2015**, *28*, 226–237. [CrossRef]

18. Pearce, J.M.; Blair, C.; Laciak, K.J.; Andrews, R.; Nosrat, A.; Zelenika-Zovko, I. 3-D Printing of Open Source Appropriate Technologies for Self-Directed Sustainable Development. *J. Sustain. Dev.* **2010**, *3*, 17–29. [CrossRef]

19. Birtchnell, T.; Hoyle, W. *3D Printing for Development in the Global South: The 3D4D Challenge*; Springer: Berlin, Germany, 2014.

20. Wittbrodt, B.; Glover, A.; Laureto, J.; Anzalone, G.; Oppliger, D.; Irwin, J.; Pearce, J. Life-Cycle Economic Analysis of Distributed Manufacturing with Open-Source 3-D Printers. *Mechatronics* **2013**, *23*, 713–726. [CrossRef]

21. Petersen, E.; Pearce, J. Emergence of Home Manufacturing in the Developed World: Return on Investment for Open-Source 3-D Printers. *Technologies* **2017**, *5*, 7. [CrossRef]

22. Kelleher, K. Was 3D Printing Just a Passing Fad? *Time* **2015**. Available online: time.com/3916323/3d-printer-stocks (accessed on 10 June 2015).

23. Bilton, R. 3D Printing Is a Gimmick, Says Foxconn Prez (and He's Sorta Right). 2013. Available online: https://venturebeat.com/2013/06/26/3d-printing-is-a-gimmick-says-foxconn-prez-and-hes-sorta-right/ (accessed on 1 July 2017).

24. Statistica. Average Amount Spent per Child on Toys by Country in 2013 (in U.S. Dollars). Available online: http://www.statista.com/statistics/194424/amount-spent-on-toys-per-child-by-country-since-2009/ (accessed on 1 July 2017).

25. Brandongaille. 23 Toy Industry Statistics and Trends. 2013. Available online: http://brandongaille.com/23-toy-industry-statistics-and-trends/ (accessed on 1 July 2017).

26. Toy Association. Annual U.S. Sales Data. Available online: http://www.toyassociation.org/tia/industry_facts/salesdata/industryfacts/sales_data/sales_data.aspx?hkey=6381a73a-ce46--4caf-8bc1--72b99567df1e#.WQqVLYjys2w (accessed on 1 July 2017).

27. Global Industry Analysts, Inc. Press Release: Toys and Games—A Global Strategic Business Report. 2015. Available online: http://www.strategyr.com/pressMCP-2778.asp (accessed on 1 July 2017).

28. LEGO: We Will Continue To Dominate The Global Toy Market. 2014. Available online: http://www.businessinsider.com/lego-we-will-continue-to-dominate-the-global-toy-market-2014--2 (accessed on 1 July 2017).

29. Ultimaker. Software—All platforms. Available online: https://ultimaker.com/en/products/cura-software/list (accessed on 1 July 2017).

30. Chilson, L. The Difference between ABS and PLA for 3D Printing. *ProtoParadigm* **2013**. Available online: http://www.protoparadigm.com/news-updates/the-difference-between-abs-and-pla-for-3d-printing/ (accessed on 1 July 2017).

31. Stephens, B.; Azimi, P.; El Orch, Z.; Ramos, T. Ultrafine Particle Emissions from Desktop 3D Printers. *Atmos. Environ.* **2013**, *79*, 334–339. [CrossRef]

32. Tokiwa, Y.; Calabia, B.; Ugwu, C.; Aiba, S. Biodegradability of Plastics. *Int. J. Mol. Sci.* **2009**, *10*, 3722–3742. [CrossRef] [PubMed]

33. Hoffman, T. LulzBot Mini 3D Printer. PCMAG. **2015**. Available online: http://www.pcmag.com/article2/0,2817,2476575,00.asp (accessed on 1 July 2017).

34. U.S. Energy Information Administration Independent Short-Term Energy Outlook (STEO). Available online: https://www.eia.gov/forecasts/steo/pdf/steo_full.pdf (accessed on 1 July 2017).

35. Amazon Hatchbox 3D PLA-1kg 3.00-BLK PLA 3D Printer filament Dimensional Accuracy ±0.05 Mm, 1 Kg Spool, 3.00 Mm, Black. Available online: http://www.amazon.com/HATCHBOX-3D-PLA-1KG3--00-BLK-Filament-Dimensional/dp/B00MEZE7XU (accessed on 1 July 2017).

36. Alibaba. PLA plastic pellets. Available online: https://www.alibaba.com/showroom/pla-plastic-pellets.html (accessed on 16 May 2017).
37. Baechler, C.; DeVuono, M.; Pearce, J.M. Distributed Recycling of Waste Polymer into RepRap Feedstock. *Rapid Prototyp. J.* **2013**, *19*, 118–125. [CrossRef]
38. Zhong, S.L.; Pearce, J.M. Tightening the Loop on the Circular Economy: Coupled Distributed Recycling and Manufacturing with Recyclebot and RepRap 3-D Printing. To be published.
39. Pearce, J. Quantifying the Value of Open Source Hardware Development. *Mod. Econ.* **2015**, *6*, 1–11. [CrossRef]
40. Pearce, J.M. Return on investment for open source scientific hardware development. *Sci. Public Policy* **2016**, *43*, 192–195. [CrossRef]
41. Wang, Z. Mattell: Buy the Toys While They're on Sale. 2015. Available online: http://seekingalpha.com/article/3160396-mattel-buy-the-toys-while-theyre-on-sale (accessed on 1 July 2017).
42. Allain, R. How Much Does One Lego Piece Cost? 2014. Available online: https://www.wired.com/2014/08/lego-cost/ (accessed on 1 July 2017).
43. Griepp, M. Hobby Games Market Nearly $1.2 Billion. ICv2. 2016. Available online: https://icv2.com/articles/news/view/35150/hobby-games-market-nearly-1--2-billion (accessed on 1 July 2017).
44. Wong, J.I. Old-fashioned board games, not tech, are attracting the most money on Kickstarter. 2016. Available online: https://qz.com/688843/old-fashioned-board-games-not-tech-are-attracting-the-most-money-on-kickstarter/ (accessed on 1 July 2017).
45. Business Wire. Top 3 Trends Impacting the Global Board Games Market Through 2021: Technavio. 2016. Available online: http://www.businesswire.com/news/home/20161228005057/en/Top-3-Trends-Impacting-Global-Board-Games (accessed on 1 July 2017).
46. Appropedia. Save the Planet Board Game. Available online: http://www.appropedia.org/Save_the_planet_board_game (accessed on 1 July 2017).
47. Amazon Best Sellers. Board Games. Available online: https://www.amazon.com/Best-Sellers-Toys-Games-Board/zgbs/toys-and-games/166225011 (accessed on 3 May 2017).
48. English Oxford Living Dictionaries- Cosplay. Available online: https://en.oxforddictionaries.com/definition/cosplay (accessed on 4 May 2017).
49. Seregina, A.; Weijo, H. Play at Any Cost: How Cosplayers Produce and Sustain Their Ludic Communal Consumption Experiences. *J. Consum. Res.* **2017**, *44*, 139–159. [CrossRef]
50. Le Duigou, A.; Castro, M.; Bevan, R.; Martin, N. 3D printing of wood fibre biocomposites: From mechanical to actuation functionality. *Mater. Des.* **2016**, *96*, 106–114. [CrossRef]
51. Rudnicki, M.; Pringle, A.M.; Pearce, J.M. Viability of Up-Cycling Wood Furniture Waste to 3-D Printing Filament. In *Advancements in Fiber-Polymer Composites Symposium*; 2017, in press.
52. Kreiger, M.; Pearce, J.M. Environmental impacts of distributed manufacturing from 3-D printing of polymer components and products. In *MRS Proceedings*; Cambridge University Press: Cambridge, UK, 2013; Volume 1492, pp. 85–90.
53. Kreiger, M.; Pearce, J.M. Environmental life cycle analysis of distributed three-dimensional printing and conventional manufacturing of polymer products. *ACS Sustain. Chem. Eng.* **2013**, *1*, 1511–1519. [CrossRef]
54. Kostakis, V.; Papachristou, M. Commons-based peer production and digital fabrication: The case of a RepRap-based, Lego-built 3D printing-milling machine. *Telemat. Inf.* **2014**, *31*, 434–443. [CrossRef]
55. Adapterz LLC. Lego to Lincoln Logs. Available online: https://www.myminifactory.com/object/lego-to-lincoln-logs-uck-05f06m-14924 (accessed on 5 May 2017).
56. Amazon. Building Bricks 0 Regular Colors—1000 Pieces—Compatible with All Major Brands. Available online: https://www.amazon.com/Building-Bricks-Regular-Colors-Compatible/dp/B015EQIOCA/ref=sr_1_1 (accessed on 1 July 2017).
57. STARTT 3D Printer. iMakr.com. Available online: https://www.imakr.com/us/en/startt-affordable-3d-printer/1146-startt-3d-printer.html (accessed on 1 July 2017).
58. Bak, D. Rapid prototyping or rapid production? 3D printing processes move industry towards the latter. *Assem. Autom.* **2003**, *23*, 340–345. [CrossRef]
59. Petrick, I.J.; Simpson, T.W. 3D printing disrupts manufacturing: How economies of one create new rules of competition. *Res.-Technol. Manag.* **2013**, *56*, 12–16. [CrossRef]
60. Berman, B. 3-D printing: The new industrial revolution. *Bus. Horiz.* **2012**, *55*, 155–162. [CrossRef]

61. Kietzmann, J.; Pitt, L.; Berthon, P. Disruptions, decisions, and destinations: Enter the age of 3-D printing and additive manufacturing. *Bus. Horiz.* **2015**, *58*, 209–215. [CrossRef]
62. Rifkin, J. *The Zero Marginal Cost Society: The Internet of Things, the Collaborative Commons, and the Eclipse of Capitalism*; Palgrave Macmillan: Basingstoke, UK, 2014.
63. Laplume, A.O.; Petersen, B.; Pearce, J.M. Global value chains from a 3D printing perspective. *J. Int. Bus. Stud.* **2016**, *47*, 595–609. [CrossRef]
64. Rehnberg, M.; Pointe, S. *3D Printing and Global Value Chains: How a New Technology May Restructure Global Production*; Global Production Networks Centre: Singapore, 2016; GPN2016-010; Available online: http://gpn.nus.edu.sg/file/Stefano%20Ponte_GPN2016_010.pdf (accessed on 1 July 2017).
65. Pîrjan, A.; Petrosanu, D.M. The Impact of 3D Printing Technology on the Society and Economy. *J. Inf. Syst. Oper. Manag.* **2013**. Available online: ftp://ftp.repec.org/opt/ReDIF/RePEc/rau/jisomg/Wi13/JISOM-WI13-A19.pdf (accessed on 7 July 2017).
66. Thiesse, F.; Wirth, M.; Kemper, H.G.; Moisa, M.; Morar, D.; Lasi, H.; Piller, F.; Buxmann, P.; Mortara, L.; Ford, S.; Minshall, T. Economic Implications of Additive Manufacturing and the Contribution of MIS. *Bus. Inf. Syst. Eng.* **2015**, *57*, 139. [CrossRef]
67. Aitken-Palmer, W. A Market-Based Approach to 3d Printing for Economic Development in Ghana. Master's Thesis, Michigan Technological University, Houghton, MI, USA, 2015.
68. Pearce, J.M. Applications of open source 3-D printing on small farms. *Org. Farming* **2015**, *1*, 19–35. [CrossRef]
69. Pearce, J.M. Emerging Business Models for Open Source Hardware. *J. Open Hardw.* **2017**, *1*, 2. [CrossRef]
70. Reed, R.; Storrud-Barnes, S.; Jessup, L. How open innovation affects the drivers of competitive advantage: Trading the benefits of IP creation and ownership for free invention. *Manag. Decis.* **2012**, *50*, 58–73. [CrossRef]
71. Peh, Z.K.; Yap, Y.L.; Yeong, W.Y.; Liow, H.H. Application of 3D printed medical aid for pediatric cancer patients. In Proceedings of the 2nd International Conference on Progress in Additive Manufacturing, Nanyang Technological University, Singapore, 16–19 May 2016; pp. 49–54.
72. Saengchairat, N.; Tran, T.; Chua, C.K. A review: Additive manufacturing for active electronic components. *Virtual Phys. Prototyp.* **2017**, *12*, 31–46. [CrossRef]
73. Anzalone, G.C.; Wijnen, B.; Pearce, J.M. Multi-material additive and subtractive prosumer digital fabrication with a free and open-source convertible delta RepRap 3-D printer. *Rapid Prototyp. J.* **2015**, *21*, 506–519. [CrossRef]
74. Vidimče, K.; Wang, S.P.; Ragan-Kelley, J.; Matusik, W. OpenFab: A programmable pipeline for multi-material fabrication. *ACM Trans. Graph.* **2013**, *32*, 136. [CrossRef]
75. Khoo, Z.X.; Teoh, J.E.M.; Liu, Y.; Chua, C.K.; Yang, S.; An, J.; Leong, K.F.; Yeong, W.Y. 3D printing of smart materials: A review on recent progresses in 4D printing. *Virtual Phys. Prototyp.* **2015**, *10*, 103–122. [CrossRef]
76. Cruz Sanchez, F.A.; Boudaoud, H.; Muller, L.; Camargo, M. Towards a standard experimental protocol for open source additive manufacturing: This paper proposes a benchmarking model for evaluating accuracy performance of 3D printers. *Virtual Phys. Prototyp.* **2014**, *9*, 151–167. [CrossRef]
77. Gewirtz, D. Adding a Raspberry Pi Case and a Camera to Your LulzBot Mini—Watch Video Online—Watch Latest Ultra HD 4K Videos Online. 2016. Available online: http://www.zdnet.com/article/3d-printing-hands-on-adding-a-case-and-a-camera-to-the-raspberry-pi-and-lulzbot-mini/ (accessed on 30 November 2016).
78. Printer3D. Free IP Camera Monitoring for 3D Printer with Old Webcam usb in 5min—3D Printers English French & FAQ Wanhao Duplicator D6 Monoprice Maker Ultimate & D4, D5, Duplicator 7, 2017. Available online: http://www.printer3d.one/en/forums/topic/free-ip-camera-monitoring-for-3d-printer-with-old-webcam-usb-in-5min/ (accessed on 18 March 2017).
79. Simon, J. Monitoring Your 3D Prints | 3D Universe. 2017. Available online: https://3duniverse.org/2014/01/06/monitoring-your-3d-prints/ (accessed on 18 March 2017).
80. Nuchitprasitchai, S.; Roggemann, M.; Pearce, J. Factors Effecting Real Time Optical Monitoring of Fused Filament 3-D Printing. *Prog. Addit. Manuf.* **2017**, 1–17. [CrossRef]
81. Nuchitprasitchai, S.; Roggemann, M.; Pearce, J. Three Hundred and Sixty Degree Real-Time Monitoring of 3-D Printing Using Computer Analysis of Two Camera Views. *J. Manuf. Mater. Process.* **2017**, *1*, 2. [CrossRef]
82. Nuchitprasitchai, S.; Roggemann, M.; Pearce, J. An Open Source Algorithm for Reconstruction 3-D images for Low-cost, Reliable Real-time Monitoring of FFF-based 3-D Printing. To be published.

83. Cruz, F.; Lanza, S.; Boudaoud, H.; Hoppe, S.; Camargo, M. Polymer Recycling and Additive Manufacturing in an Open Source context: Optimization of Processes and Methods. In Proceedings of the 2015 Annual International Solid Freeform Fabrication Symposium-An Additive Manufacturing Conference, Austin, TX, USA, 7–9 August 2015; pp. 10–12.

84. Rayna, T.; Striukova, L. From rapid prototyping to home fabrication: How 3D printing is changing business model innovation. *Technol. Forecast. Soc. Chang.* **2016**, *102*, 214–224. [CrossRef]

technologies

MDPI

Article

Open Source Multi-Head 3D Printer for Polymer-Metal Composite Component Manufacturing

John J. Laureto [1] and Joshua M. Pearce [1,2,*]

[1] Department of Materials Science and Engineering, Michigan Technological University, 601 M&M Building, 1400 Townsend Drive, Houghton, MI 49931-1295, USA; jjlauret@mtu.edu
[2] Department of Electrical & Computer Engineering, Michigan Technological University, 121 EERC Building, 1400 Townsend Drive, Houghton, MI 49931-1295, USA
* Correspondence: pearce@mtu.edu; Tel.: +1-906-487-1466

Academic Editors: Salvatore Brischetto, Paolo Maggiore and Carlo Giovanni Ferro
Received: 1 March 2017; Accepted: 18 April 2017; Published: 15 June 2017

Abstract: As low-cost desktop 3D printing is now dominated by free and open source self-replicating rapid prototype (RepRap) derivatives, there is an intense interest in extending the scope of potential applications to manufacturing. This study describes a manufacturing technology that enables a constrained set of polymer-metal composite components. This paper provides (1) free and open source hardware and (2) software for printing systems that achieves metal wire embedment into a polymer matrix 3D-printed part via a novel weaving and wrapping method using (3) OpenSCAD and parametric coding for customized g-code commands. Composite parts are evaluated from the technical viability of manufacturing and quality. The results show that utilizing a multi-polymer head system for multi-component manufacturing reduces manufacturing time and reduces the embodied energy of manufacturing. Finally, it is concluded that an open source software and hardware tool chain can provide low-cost industrial manufacturing of complex metal-polymer composite-based products.

Keywords: open source; 3D printing; RepRap; composite; manufacturing

1. Introduction

The increased utilization [1,2] of self-replicating rapid prototyper (RepRap) 3D printers [3,4] using fused filament fabrication (FFF) (material extrusions by ASTM F2792-12a: Standard Terminology for Additive Manufacturing Technologies) [5] has increased the engineering applications of polymer extrusion materials. Printable polymer material characterization has increased the knowledge available to engineers for common PLA and ABS materials [6–9] along with an increasing list of thermoplastics [10,11], polymer metal composite materials [12–14] and polymer ceramic composite materials [15–18] for a number of novel applications, including medical and health-related components [19–23]. Subsequently, advancements in material understanding has led to the development of more sophisticated RepRap machines. Currently, multi-head printers (typically two hot ends) are readily available from re:3D, Aleph Objects, Prusa Research and other open source 3D printer manufacturers, and distributed designs are downloadable with creative commons and GPL licenses from the RepRap wiki and Internet repositories of 3D designs. Multi-head printers allow for multi-color printing to achieve aesthetic requirements and/or multi-material manufacturing of the same work piece [24]. Commonly, a sacrificial material (e.g., polyvinyl alcohol) is utilized as a supporting material to be easily removed during post-processing [25]. Recently, Ma et al. developed processing techniques to manufacture heterogeneous structures/composites using thin wall mold cavities and reusable multipart molds by combining shape deposition manufacturing (SDM), FFF and casting [26]. Furthermore, while still in the early stages of development, metal printing RepRap's provide a partial step towards full adoption of additive manufacturing techniques [27] and multi-material selection in 3D manufacturing [28–32] to accommodate future requirements of material quality, design for manufacturing, processing monitor and achievement of near net shape [33]. Further expanding the RepRap machine customization is the advent of Franklin [34], an open-source 3D printing control software. Franklin's application to a variety of RepRap applications has been shown including: laser welding of HDPE polymer sheet [35], multi-material additive and subtractive fabrication [36], printed components for small organic farms [37] and voltage monitoring of a GMAW (gas metal arc welding) metal-based RepRap Delta printer [38]. Multi-material 3-D printers including those able to fabricate with composite materials such as fiber-reinforced polymer materials have been academically researched by Quan et al. [39]. Furthermore, similar to the application to be described are numerous applications of metal wire embedment into a primarily polymer matrix [40]. Recent investigative research has provided insight to copper wire encapsulation of copper for electronic sensing [41], tool path planning for wire embedment on FFF printed curved surfaces [42], metal fiber encapsulation for electromechanical robotic components [43], flexible printed circuit boards (PCB) for structural electronic devices [44] and open-source 3D printing CAD/CAM software for quality function deployment (QFD) and theory of inventive problem solving (TRIZ) optimization [45].

To further the scope of potential applications of RepRap manufacturing, this paper aims to describe a manufacturing technology that accomplishes a partial step forward to true multi-material selection. This paper provides free and open source hardware and software for printing systems that achieves metal wire embedment into a polymer matrix 3D printed part via weaving and wrapping procedures. In addition, a method utilizing OpenSCAD and parametric coding is provided that enables customized g-code commands to be developed for a given component design and material selection. Then, upon fixture placement, this method enables weaving and wrapping procedures by g-code line entries after each successive polymer layer deposition to create metal matrix composites. These composite parts are then evaluated from the technical viability of manufacturing and quality. Specifically, to identify the advantages of utilizing a multi-polymer head system for multi-component manufacturing, time studies are to be conducted and compared to traditional single-head per material manufacture of the same part. In addition, the metal/polymer interface bond strength is quantified with a burst pressure measurement. The results are presented and discussed in the context of low-cost distributed manufacturing of complex metal-polymer composites.

2. Materials and Methods

2.1. Fabrication of the Gigabot for Multi-Head Metal-Polymer Composite Printing

A re:3D Gigabot 3.0 3D printer [46] was modified for the development of the metal polymer matrix apparatus. The printing system utilizes a gantry system to accommodate five extruder nozzles and x-axis directional commands. A single NEMA 17 stepper motor with 20 tooth GT2 pulleys controls the movement of the x-axis. The y-axis commands are controlled by two NEMA 17s, one at each end of the gantry length. Similar to the y-axis, the z-axis movement is controlled by two NEMA 17s at opposite sides of the 60 cm × 60 cm (XY) build platform. Both z-axis and y-axis commands are sent to a NEMA 17 and replicated by the "follower" second motor based on the provided g-code. The printer is constructed with 80-20 extruded aluminum with bolts, nuts, fittings, threaded rods and brackets where required following the re:3D standard design. Figure 1 pictorially describes the printing apparatus to be discussed. Described are the relative locations of extruder/directional motors along with hot end locations on the x-axis gantry and electrical control board mounting locations.

Plastic 3D-printed components needed for the assembly are shown in Table 1. They were obtained through Thingiverse, a collaborative online maker space with downloadable component files (indicated by thing number in Table 1) or custom designed in OpenSCAD [47], a parameter modeling computer-aided design (CAD) software. Designed or downloaded part files were printed with polylactic acid (PLA) on either a MOST delta RepRap or a Lulzbot 5.0. Component design, coding and printing parameters allowed for easy modification, development, decreased print time and economical use of filament material. All part files (.scad/STL) (Table 1) designed by MOST in OpenSCAD are available for download [48] under the GNU GPLv3 [49]. Secondly, the complete bill of materials including metric type accessory components and electrical components is displayed in Tables 2 and 3, respectively. Operational and installation instructions are available online at Appropedia [50].

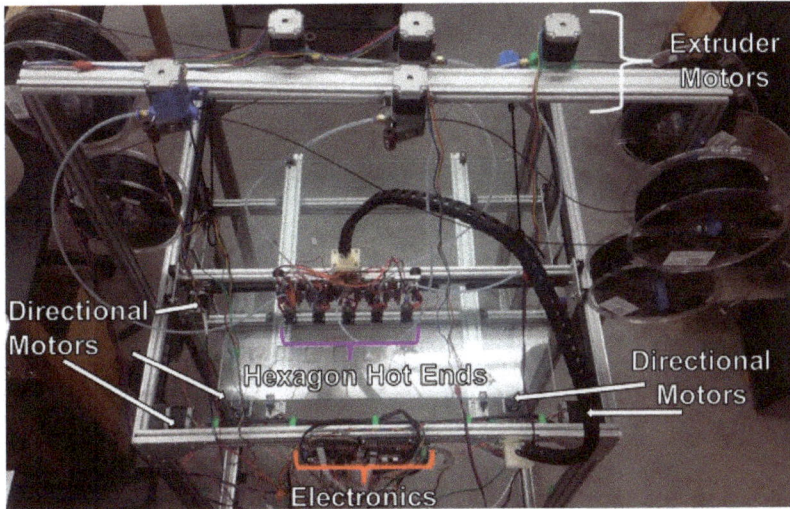

Figure 1. Complete manufactured metal-polymer composite Gigabot. Primary electromechanical components and their respective mounting locations are identified.

Table 1. Metal-polymer composite Gigabot printed/structural components.

Part Name/Description	Count	Rendered Image	Part Name/Description	Count	Rendered Image
Extruder Mount Bracket	5		z-Height End Stop Solenoid Mount	2	
z-Height Bed Leveling Adjustment	5		80-20 Wire Guides	10	
z-Height Bed Leveling Dovetail Mounts	5		Gantry Cable Supports	3	
Filament Spool Holders thing:1269563	6		Build Plate Fixturing Brackets	4	
80-20 M4 T-Slot Mount thing:1061769	2		Hexagon Hot end Fan Mount	5	

Table 1. *Cont.*

Part Name/Description	Count	Rendered Image	Part Name/Description	Count	Rendered Image
z-Height Z1 and Z0 Leveling Screw Knob	2		MOST Bowden Extruder Drive	5	
Gantry Mount Cable Carrier Connection	1		Arduino Mega 250 Mount Bracket	1	
Customized I/O Board Mount Bracket	1		80-20 Cable Carrier Mount	1	
Gantry Mount Electrical Connection Board	1		y–Carriage Belt Clamp	2	
y-Axis End Stop Solenoid Mount	2		Compact Bowden Extruder thing:275593	1	

Table 2. Metal-polymer composite Gigabot mechanical bill of materials.

Part Description	Count	Source	Serial/Pat Number
GT2 3MR 9-mm Wide	1(15 ft)	Gates	-
GT2 Timing Pulley	3	Gates	-
9-mm Idler Pulley with 625-2RS Bearings	3	re:3D	-
M5x8 Button Head Cap Screws	100	BoltDepot.com	-
Hexagon Full Metal Hot-End 1.75 mm, 12 V	5	IC3D–Hexagon	X000SV0T0N
Cyclemore 1.0-mm Brass Nozzle	5	Cyclemore	X000WJAXH5
PC4-M6 Push-In Fitting	10	Cyclemore	30-60007-016-FBA
53 Link Cable Carrier	1	Re:3D	-
Teflon (PTFE) Bowden Tube 1.75 mm (2.0 mm ID/4.0 mm OD)	25 ft	3D CAM	BOWDEN2M
3/8"–8 ACME Threaded Rods	2	re:3D	-
V-Grove Roller Bearings	20	Re:3D	-
67 mm × 60 mm Annealed Glass Build Plate	1	Locally sourced	-
80-20 Series 20 T-Slot Nuts	100	re:3D	-
Threaded Rod Z-Nut Cup	2	re:3D	-
MXL 18 Tooth Motor Pulley	2	re:3D	-
MXL 36 Tooth Motor Pulley (Threaded Rod)	2	re:3D	-
z-Axis MXL Belt	2	re:3D	-
Aluminum Side Plate	4	re:3D	-
Aluminum Corner Plate	8	re:3D	-
Rectangular Brackets for Extruder Motor Gantry	2	re:3D	-
3 mm × 9 mm Stainless 18-8 Washer	100	BoltDepot.com	7319
DIY: Gigabot Parts Kit	1	re:3D	-
M2 Hex Nut	100	BoltDepot.com	-
Eccentric Wheel Spacer	4	re:3D	-
Z-Motor Shelf	2	re:3D	-
Truck Plates (L/R)	2	re:3D	-
Thermal Tape	10	adafruit	1468
A4988 Pololu Heat Sink	10	Pololu Robotics and Electronics	-

Table 3. Metal-polymer composite Gigabot electrical components.

Part Description	Count	Source
NEMA 17 Stepper Motor	10	-
RAMPS 1.4	2	-
A4988 Pololu Driver	10	-
Arduino 250 Mega	1	-
Custom I/O Board	1	-
12 V Power Supply	1	-
36 V Power Supply	1	-
End Stop Solenoid Limit Switches	5	re:3D

The x-axis gantry is installed with five full metal 1.75-mm hexagon hot ends [51] spaced 55 mm apart. Spacing of the hot-end is controlled by two manufactured aluminum plate measuring 3.175 mm × 25.400 mm × 295.75 mm. The 55mm spacing is driven by the placement of the z- leveling dovetail mounting points. The aluminum plates and z-leveling dovetails are fixtured by the application of M5 bolts and roller bearings. The printed hexagon mounting fixture is a tongue and groove design allowing for independent z-axis leveling with adjustment of an M3 set screw, i.e., each extruder nozzle is individually leveled to the build platform. This allows for replicate parts to be simultaneously printed assuming that gcode commands do no exceed the 55mm spacing machine constraint. Figure 2 displays the x-gantry mounting system.

Figure 2. X-axis gantry assembly. 5× Hexagon Full-Metal 12 V hot ends are shown fixtured to their respective 'Z-Height Bed Leveling Adjustment' part files. As shown, dovetail leveling mechanisms are attached to the machined aluminum plate (3.175 mm × 25.4 mm × 295.75 mm) with Hexagon nozzle diameter cylindrical axis spaced 55 mm.

The five hexagon hot ends are provided filament through Bowden sheaths constructed from 4 mm OD (2 mm ID) pressure fitting compatible polytetrafluoroethylene (PTFE) flexible tubing. The Bowden extruder system decreases the weight on the x-axis gantry, thus allowing for faster and more accurate prints. Decreased weight on the x-axis gantry is also advantageous, as it will decrease the likely hood of the single x-axis NEMA 17 skipping, leading to a loss of positioning. The Bowden extruder bodies, NEMA 17s and assembly structures are mounted to the secondary elevated gantry. M5 and t-slot nuts allow proper fixture to the secondary 80–20 aluminum gantry. Figure 3 provides further details of the five extruder motors installed on the gantry along with a close up image of the extruder motor assembly. Furthermore, the feed filament is spooled adjacent to its respective extruder motor.

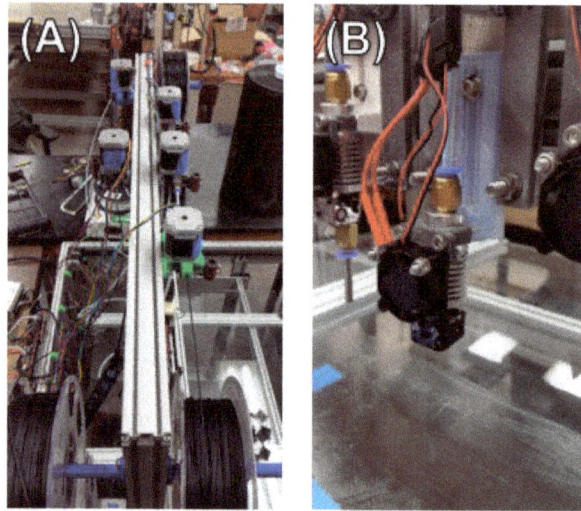

Figure 3. (**A**) Top printer gantry with fixtured 5 NEMA 17 extruder drive motors and respective "MOST Bowden Extruder Drive" printed components; (**B**) hexagon hot-end assembly detail with "Z-Height Bed Leveling Adjustment" dovetails.

Additionally, due to the large build platform, two z-axis zeroing locations are utilized. Two M5 screws with fitted ergonomic adjustment knobs and tension springs allow for z-axis leveling independently. Upon proper adjustment, the x-axis gantry extruders can be leveled to the build platform. Figure 4 displays the leveling system.

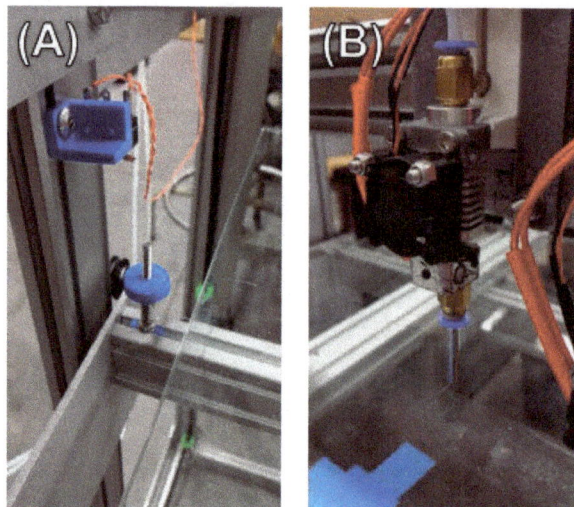

Figure 4. (**A**) Bed platform Z-height leveling. Shown are "Z-Height End Stop Solenoid Mount", "80-20 M4 T-Slot Mount thing: 1061769" and "Z-Height Z1 and Z0 Leveling Screw Knob" fixtured to 80-20 aluminum rails with M5 nuts; (**B**) height adjustment assembly shown at the maximum height adjustment in contrast to Figure 3B.

2.2. Circuit Assembly and Printer Control

To accommodate the quantity of NEMA 17 stepper motors, solenoid end stops and thermistors, a custom circuit board enabling the application of two RAMPS (RepRap Arduino Mega Pololu Shield) 1.4 is created [52]. Application of this circuit, as described in Figures 5 and 6, provides two functional RAMPS 1.4 and subsequent A4988 stepper motor driver carriers [53] from one Arduino Mega 2560 [54]. The KiCad-PcbNew 4.0.3 designed I/O board communicates with the secondary RAMPS 1.4 board allowing for the further allocation of pins on the Arduino microcontroller [55]. Pin assignments, as presented in the Franklin printer profile, are shown in Tables 4 and 5. The A4988 potentiometers are adjusted to provide 0.6–1.2 mV of potential measured between ground. Each potentiometer is fitted with an aluminum heatsink fixture with thermal tape to aid in temperature control.

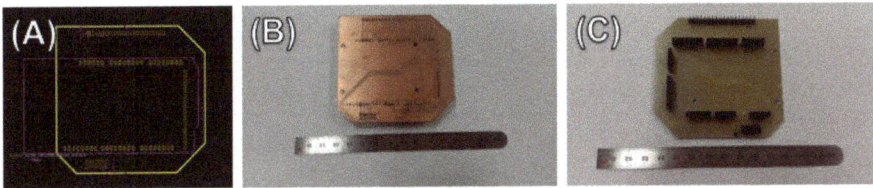

Figure 5. (**A**) Electrical diagram/schematic developed in KiCAD-PcbNew; (**B**) milled PCB surface for representation; (**C**) PCB pin side for representation.

Figure 6. Assembled two RAMPS 1.4 with custom I/O PCB per the KiCAD-PcbNew specification.

Table 4. Stepper motor pin assignments [1].

Pin Type	X_D	$Y0_D$	$Y1_D$	$Z0_D$	$Z1_D$	$Ex0_E$	$Ex1_E$	$Ex2_E$	$Ex3_E$	A_E
Step	D32	D60(A6)	D43	D46	D37	D29	D36	D26	D54(A0)	D35
Direction	D47	D61(A7)	D41	D48	D39	D31	D34	D28	D55(A1)	D33
Enable	D45	D56(A2)	D45	D62(A8)	D45	D45	D30	D24	D38	D45
Min Limit	D3	D14	D23	D18	D0	D0	D0	D0	D0	D0
Max Limit	D2	D15	D25	D19	D0	D0	D0	D0	D0	D0

[1] Pin assignments are relative to the A4988 and stepper motors physical location on the RAMPS 1.4. Refer to Figure 5 for specific location details.

As indicated in Table 5, 24-V heater cartridges, cooling fans and thermistors are connect to their respective RAMPS 1.4 positions through a secondary custom I/O board. The I/O board acts as a central hub for all communication to the components on the x-axis gantry. Figure 7 identifies the location of this board and the connection points of each component, while Figure 8 describes the PCB in greater detail.

Table 5. Hexagon hot end Arduino pin assignments.

Pin Type	$Ex0_E$	$Ex1_E$	$Ex2_E$	$Ex3_E$	A_E
Heater	D9	D10	D42	D64(A11)	D8
Fan	D0	D0	D0	D0	D0
Thermistor	A14(D68)	A15(D69)	A10(D64)	A12(D66)	A13(D67)

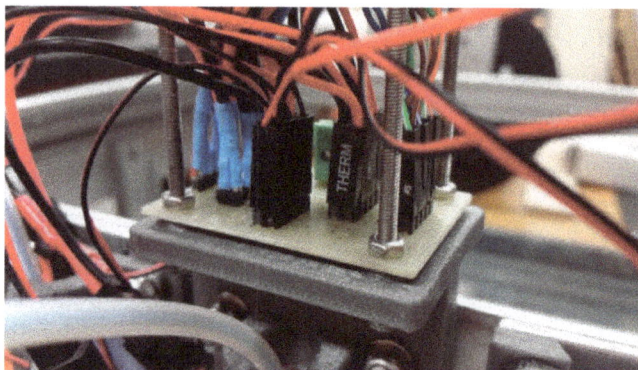

Figure 7. Assembled secondary I/O PCB for x-axis gantry components.

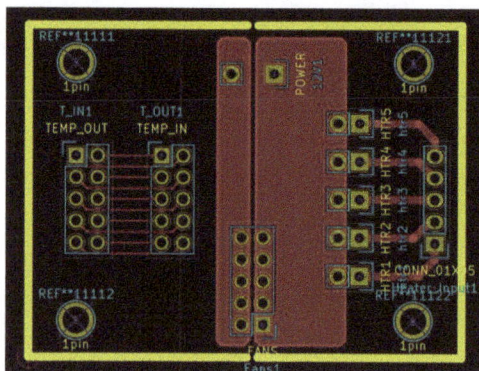

Figure 8. Secondary I/O PCB schematic developed in KiCAD-PcBNew. Connection zones as indicate in this image are further indicated in Figure 7 as previously described.

The metal-polymer composite Gigabot requires two power supplies to meet full operational requirements. As designed, an input 110/220 V, output 12 V 20 A power supply is utilized for thermistor operation. An input: 110/220 V, output: 36 V 10 A power supply enables the operation of both RAMPS 1.4 boards and the secondary custom I/O board. Thus, location, position and extruder motor(s) operate on a separate power supply as compared to the thermistors and heater cartridges.

In total, ten NEMA 17 motors need to be controlled for proper functionality of the printer assembly. Specifically, there is a NEMA 17 assigned to each movement axis as listed; X, Y0, Y1, Z0, Z1, E0, E1, E2, E3 and A. Further functional description of each motor is shown in Table 6 along with a qualitative electromechanical process map, shown in Figure 9, indicating primary connection mechanisms' hot ends, thermistors, heater cartridges, end stops, extruder motors and directional motors.

Table 6. NEMA 17 motor settings and physical description.

Motor (x_D = direction, x_e = extruder)	Coupling (steps/mm)	Limit Velocity (mm/s)	Limit Acceleration (mm^2/s)
X_D	59.292	150	250
$Y0_D$	59.292	150	250
$Y1_D$	59.292	150	250
$Z0_D$	2133.333	4	250
$Z1_D$	2133.333	4	250
$Ex0_E$	100	200	1000
$Ex1_E$	100	200	1000
$Ex2_E$	100	200	1000
$Ex3_E$	100	200	1000
A_E	100	200	1000

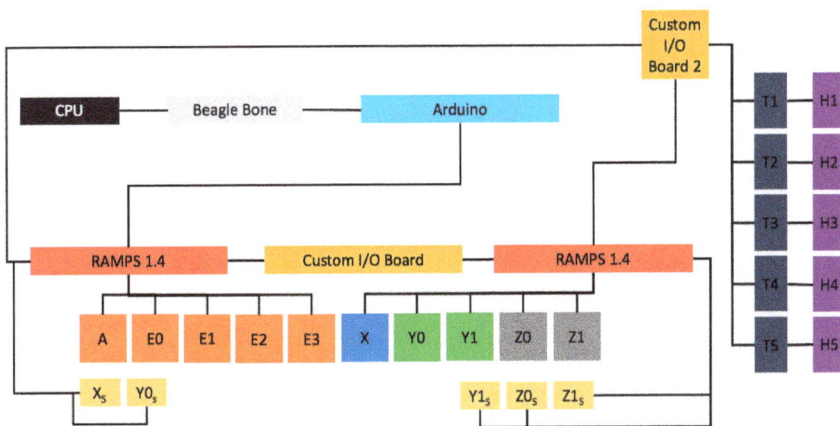

Figure 9. Electromechanical process map of the metal-polymer composite Gigabot. This diagram represents a qualitative understanding of the primary connection points between operational mechanisms and electronic controllers. Extruder motors: A, E0, E1, E2, E3. Directional motors: X, Y0, Y1, Z0, Z1. Solenoid end stops: X_S, $Y0_S$, $Y1_S$, $Z0_S$, $Z1_S$. Thermistors: T1, T2, T3, T4, T5. Heater cartridges: H1, H2, H3, H4, H5.

The open-source firmware (Franklin) controls the motion of the printer assembly. The graphic user interface (GUI) of Franklin provides the user with an interface in which to upload g-code and customize printer settings and parameters. g-code and printer settings are communicated to the printer through the host computer into to the controller. Respective g-code is formulated upon the generation of a stereolithography file (e.g., STL file). Print layer g-code was developed with Slic3r 1.2.9 [56]. The resultant g-code is typical such that the application into any RepRap printer should be easily achieved. Unique, however, is the metal-polymer composite Gigabot's multiple motors per axis (e.g., Y0/Y1, Z0/Z1 and Ex0/Ex1/Ex2/Ex3). In the current state, Slic3r in unable to individually command multiple extruders and axis motors simultaneously. Subsequently, Franklin allows for motors to be controlled via a "leader and follower" principle. For example, a printer controlled by Franklin a g-code command of "G1 Y213 Z55" will signal movement of Y0/Y1 and Z0/Z1 to a relative position of 213 mm and 55 mm, respectively. In effect, the g-code command pulsed through the controller to the stepper motor is initially recognized by the "leader" (i.e., Y0 or Z0 and henceforth followed and/or replicated by Y1 or Z1). The resultant interaction is duplicate movements by the affected stepper motors. The "leader and follower" principle are also used for the Ex0E-Ex3E extruder motors (i.e., four of the five hot ends will extrude the same portion of filament based on a standard g-code command). In this circumstance, Ex0E is the leader extruder followed by Ex1E, Ex2E and Ex3E. Unique to the metal-polymer composite Gigabot machine is extruder AE. Functionally, AE, is a directional movement axis, which has been modified to be used as an extruder. The proper coupling, limit velocity and acceleration settings in Franklin allow for this change. Separation of AE from Ex0E–Ex3E allows for individualized commands within the g-code. Other than "E" commands, Slic3r cannot currently generate extruder commands for different extruders. To introduce "A" commands, visual basic applications were utilized to reformat the text of the outputted g-code. Table 7 describes a sample operation of this process.

Table 7. Visual Basic g-code modifier (spreadsheet reference cell#).

	Initial g-code Command Line	G1 F900 X143.487 Y114.988 E0.51434 (A27)
Operation 1	=IF(ISNUMBER(SEARCH("G1",A27)),RIGHT(A27,LEN (A27)-SEARCH("E",A27,1)+1),"NA")	E0.51434
Operation 2	=IF(ISNUMBER(SEARCH("G1",A27)),RIGHT(A27,LEN (A27)-SEARCH("E",A27,1))," ")	0.51434(E27)
Operation 3	=IF(ISNUMBER(SEARCH(" ",E27))," ","A")	A(F27)
Operation 4	=CONCATENATE(F27,E27)	A0.51434(G27)
Operation 5	=IFERROR(IF(ISNUMBER(SEARCH(" ",G27)), A27,CONCATENATE(A27&" "&G27)),A27)	G1 F900 X143.487 Y114.988 E.51434 A0.51434

The process described in Table 7 is for the utilization of all five hot ends for replicate polymer component printing. However, there are applications in which AE may be used independently relative to Ex0E–Ex3E. In these unique circumstances, g-code for AE is made separately and then superimposed on the g-code for Ex0E–Ex3E, resulting in a composite g-code.

2.3. Modification of Extruder A_E for Wire-Feeding

A modified Bowden extruder design (thing: 275593) was utilized for a wire feeding/guide apparatus. The print assembly and miscellaneous hardware were assembled in a standard manner; however, the MK7 drive gear was inverted. Inversion of the MK7 drive gears allows for a smooth, non-galled, surface to contact the metal wire. Electrical tape surface coatings were applied to both the 608zz idler bearing and the smooth end of the MK7 drive gear for the grip of a wire. The feed wire spool is mounted near the wire extruder such that the top dead center is tangent to the primary axis of the Bowden feed pathway. Figure 10 displays the assembled metal wire feeder. Utilizing the same Bowden sheath as would a polymer filament, an 1100 series aluminum wire with a diameter of 0.508 mm $\pm$ 0.012 mm is directed down through a Hexagon hot end. In a modified application such as this, the hexagon hot end nozzle has been removed while the main assembly is present to help guide the wire. A M5 pressure fitting, similar to those in the Bowden sheath assembly, is mounted to the hot end in replacement of the 1.0-mm nozzle. The utilized pressuring fitting allows for installation of a 304 stainless tube with an outer-diameter (OD) of 1.422 mm, (-0.050 mm to $+0.101$ mm) and inner-diameter (ID) of 2.184 mm. The outer diameter is equivalent to a standard 4 mm (OD) and secures properly into a M5 pressure fitting. The ID is substantial enough to allow for passage of the 0.508 mm diameter wire while also providing room for a PTFE fitting to decrease wire friction while the wire exits the tube. The wire feed guide tube and remaining extruders (Ex0E–Ex3E) are run simultaneously. Thus, the 304 tubing prior to installation in the pressure fitting is cut to a length of ~46 mm. Thus, all extruder nozzles and wire guide tubes can be leveled to the build platform at a similar height. Figure 11 displays the assembly of the structure.

Figure 10. Wire feed Bowden assembly assembled with supplementary hardware and "Compact Bowden Extruder thing: 275593". Note that a common 1.75 mm extruder drive tooth gear has been inverted and coupled with electrical tape to provide frictional rolling resistance to aid in guiding the 1100 series aluminum wire.

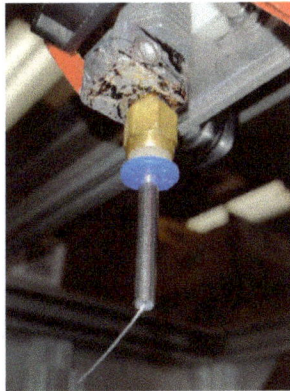

Figure 11. Wire feed guide tube. As shown, a standard M6 4.0 mm press fitting accommodates the standard threading in the hot-zone of a standard hexagon 12 V hot end. 304 stainless tubing press fits similar to a 4.0 mm PTFE tubing. Scrap PTFE tubing is fixtured to the exit zone of the 304 stainless tubing to reduce the friction associated with wire wrapping processes.

The wire feed Bowden assembly enables the ability for small increments (e.g., 1–10 mm) of wire feeding based on an AE g-code command. However, the drive mechanism is not primarily responsible for the displacement of aluminum wire. In practice, an initial length of wire is fed through the wire guide. The excess length is fixed to a pin located on the metal-polymer composite Gigabot's build platform. Controlled movement of the wire feed cross head allows for accurate positioning of the aluminum wire. As shown in Figure 9, placement of multiple secondary pins will allow for wrapping of the aluminum wire. Positioning of the fixture on the build plate is critical to the success of the wrapping procedure. Secondary fixtures are independent of the metal-polymer composite Gigabot's motor controllers; thus, offsets, in Slic3r, are to be programmed into resultant g-code. The offsets are readily determinable by manually progressing the wire-feed hot end to a known location on the secondary fixture and recording the positional coordinates provided by Franklin's GUI. The deviation in positional coordinates between the known location on the secondary fixture and Franklin's GUI output corresponds to the offsets required. In this application, positioning is only critical and programmable in the two-dimensional (XY) realm, as the z-axis, as mentioned earlier, is adjusted mechanically by the operator. The primary g-code, responsible for the wire rapping operations, can be produced from a digital parametric model. In this method, the model is set up to accommodate the fixture as shown in Figure 12. For proper generation of both the fixture and parametric wrapping, the model must be modeled in the same relative positioning. In these analyses, OpenSCAD modeling was used to model the entire print pre-production. Figure 13 displays a rendering of the OpenSCAD modeling.

Figure 12. In situ process photo of Franklin-controlled wire wrapping.

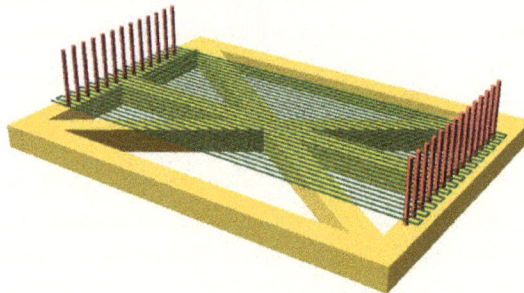

Figure 13. Rendering of parametric OpenSCAD model (yellow: fixture; red: pins; green: wire). In situ process of the designed OpenSCAD model displayed in Figure 12.

g-code generated for the wire weaving is obtained by individually exporting the (green) wire portion as an STL. The exported model can be placed into Slic3r and sliced into a single layer forgoing any Z-components. The fixture, pins and wire modeling in OpenSCAD all share an equivalent "zero" position. Thus, assuming that a specific location on the fixture can be located, the required offsets to realign the digital wire model to the physical fixture can be obtained. Typically, modifications to the generated wire wrapping g-code are required, as the models shown in Figure 11 are designed to a nominal dimension. Thus, no tolerance is designed for accommodating manufacturing/assembly of fixture positioning issues. The total realized errors, due to assembly accuracy, are not realized until initial test prints begin.

2.4. Composite Printing-Utilizing Wire-Feed Guide and Standard Brass 1.0 mm Extruder Nozzle

Slic3r 1.2.9 allows for the placement of custom g-code before and/or after a layer has been completed. Application of this software utility allows for customized wire weaving operations to occur during a standard print operation. Thus, composite structures containing 1100 series aluminum wire along with polymer FFF materials are realized. Developed processing parameters, metal-polymer composite printer modifications are all in an effort to accommodate pre-prescribed models relative to the funding agency project scope. Further secondary operations during printing are required. For example, the aluminum wires need to be heated to an elevated temperature such that the localized polymer material, at each intersection, is melted. Currently, a heater is utilized to elevate the local temperature of the metal/polymer interface. The localized heating enables the 1100 series aluminum wire and polymer material to bond sufficiently and provide significant z-height clearance for the subsequent layers of polymer material.

2.5. Polymer Filament Material Selection and Printing Parameter Development

Readily available polymer materials polyethylene terephthalate glycol modified (PETG) and polypropylene (PP) were selected for analysis. PETG was sourced from Shenzhen Esun Industries Co., Ltd. (eSUN, Shenzhen, China) and the PP from Gizmo Dorks (Temple City, USA). Materials were procured in 1 kg filament spools with a nominal diameter 1.75 mm $\pm$ 0.05 mm where roundness tolerances were not considered. Relevant intrinsic materials properties, as described by the respective materials' technical data sheets, are displayed in Table 8 [57,58].

Table 8. Material properties of PETG and PP.

-	eSUN PETG	Gizmo Dorks PP [1]
Print Temperature (°C)	230–250	230–260
Build Plate Temperature (°C)	80 or none	60
Feeding Speed (mm/s)	30–80	90

[1] Gizmo Dorks presents further and more detailed parameter settings beyond those presented here.

A variety of experimental trial prints and manufacturing runs were conducted to optimize the printing parameters. The primary metrics considered include: print speed (mm/s), extrusion/hot end temperature (°C), layer height (mm), nozzle diameter (mm), shell thickness (mm) and bottom/top layer thickness (mm). An optimized parameter set yields a quality component upon visual inspection and can be quantified with interface adhesion. Developed parameter sets are discussed and further evaluated below.

2.6. Composite Printing Tests for Metal-Polymer Composite Gigabot

Test coupons were generated using OpenSCAD to dimensions of 25.4 mm × 25.4 mm × 25.4 mm. The coupon geometry was selected to provide a simplistic volumetric model for which to compare print quality and to provide power consumption data. Print quality was determined by metrics quantifiable by visual inspection and digital caliper measurements ($\pm$ 0.01 mm) (e.g., surface smoothness, dimensional accuracy and apparent layer adhesion). Dimensional adherence to the as-designed nominal dimension of the test coupon is deeply dependent on the sliced parameter set. The intent of the dimensional analysis is to quantify the part dimensional stability per extruder, not to determine the optimum parameter set to produce nominal and/or accurate components (i.e., $\pm$ 0.127 mm). Energy consumption measurements were performed with a multi-meter for cumulative kWh monitor ($\pm$ 0.01 kWh) and instantaneous power draw (Watts).

3. Results

3.1. Resultant Print Quality and Power Consumption Measurements

An example of the resultant polymer-metal composite structure is shown in Figure 14.

Figure 14. Metal polymer composite generation dimensionally accurate to prescribed models. Cross-flow media is 1100 series aluminum wire, encased in a polymeric matrix of PETG.

Resultant print quality is shown in Figure 15. A layer height of 0.5 mm was utilized in conjunction with a 1.0-mm brass nozzle. Evidence of the relatively large layer height and nozzle are shown on the component surfaces. Wave patterns apparent on the exterior perimeters of the test coupon(s) are the result of the twenty-five percent infill percentage parameter. Wave "peaks" are adjacent to vector pathways of the infill section lines on the interior surface of the perimeter. Dimensional measurements identifying deviation from nominal are shown in Table 9. Width, length and height correspond primarily to the x, y and z coordinates, respectively.

Figure 15. Printed component part quality (visual inspection) prior to removal from substrate to be measured for dimensional precision. As indicated in Table 9, dimensional variation between hot ends is determined to be a critical metric in contrast to print parameter adjustable deviation from nominal dimensional values (i.e., 25.4 mm).

Electrical power draw (Watts) for a variety of operating conditions is shown in Table 10. Conditions were selected to identify the power requirements for each component of the metal-polymer composite Gigabot, including thermistors, heater cartridges and stepper motors (extruder and position).

Table 9. Printed component average dimensions relative to nominal dimensions (±mm).

Dimension	(A_E)	±σ	$(E0_E)$	±σ	$(E1_E)$	±σ	$(E2_E)$	±σ	$(E3_E)$	±σ
X (Width) [1]	25.82	0.08	25.86	0.08	25.93	0.06	25.94	0.06	25.93	0.18
Y (Length) [1]	25.77	0.06	25.80	0.04	25.79	0.06	25.88	0.01	25.70	0.04
Z (Height) [1]	26.42	0.09	26.66	0.05	26.39	0.06	26.45	0.04	26.25	0.03

[1] Nominal designed dimension of 25.4 mm. Averages determined from a sample size of three measurements.

Table 10. Power consumption for various metal-polymer composite Gigabot operating conditions (Watts).

Operating Condition	Power Draw (Watts) [1]
36 V 10 A Stand Alone	5.9–6.9
12 V 20 A Stand Alone	13.2–14.0
12 V 20 A with Heaters On	117–118
12 V 20 A and 36 V 10 A with Heaters On	138–144
12 V 20 A and 36 V 10 A Temp Limit (220 °C)	138–144
12 V 20 A and 36 V 10 A with Motors Enabled	45.9–46.7
12 V 20 A and 36 V 10 A with no Heaters or Motors	21–22
12 V 20 A and 26 V 10 A Motors on Heaters on and Printing	138–144

[1] Measurements are recorded in an enabled state, but idle condition, i.e., not performing a build sequence

Cumulative kWh, per print cycle, measurements are displayed in Table 11. Four parameter sets were utilized for this analysis utilizing the same test coupon geometry to quantify visual part quality. The four conditions were setup as follows: twenty-five percent infill ×5 extruders, one-hundred percent infill ×5 extruders, twenty-five percent infill ×1 extruder and 100 percent infill ×1 extruder. Single extruder studies used the A_E stepper motors and respective heater elements to print five test coupons. Conversely, multi-extruders utilized five extruders' replication the actions of A_E. The metal polymer composite Gigabot was allowed four minutes of heat up from 100 to 220 °C for each condition. All print cycles resulted in five printed components.

Table 11. Energy consumption (kWh) measurements for various print cycles.

Conditions	Metrics	Heat Up	Build	Total kWh
25% In-Fill and ×5 Extruders	Time (min)	4	9	0.03
	Cycle Power (Watts) [1]	140–144	138–144	
100% In-Fill and 5× Extruders	Time (min)	4	20	0.06
	Cycle Power (Watts) [1]	140–144	138–144	
25% In-Fill and 1× Extruders	Time (min)	4	10	0.01
	Cycle Power (Watts) [1]	68–69	62–65	
100% In-Fill and 1× Extruders	Time (min)	4	21	0.03
	Cycle Power (Watts) [1]	68–69	62–65	

[1] Table 10 measured values.

3.2. Printing Parameter and Material Development

Slic3r 1.2.9 was selected as the primary slicing tool for g-code generation. As compared to Cura 15.04.6, Slic3r allowed for custom g-code, including: start g-code, end g-code, before layer change g-code and after layer change g-code [59]. Without the implementation of this interface combining metal wire wrapping processes with the polymer, printing would not be possible.

PETG was selected as the primary polymer material for the metal/polymer composite over PP. In virgin filament form, PETG is rigid in comparison to PP. During manufacturing trials, PP would consistently twist and bend within the Bowden sheath, thus causing filament jams.

Developing processing techniques to ensure consistent material flow throughout the hot end was troublesome. Secondly, PP requires like to like material for proper build plate adhesion. Specifically, PP build plates are required to reduce delamination part warping after deposition. Conversely, PETG is readily suited to adequately bond to a glass build plate with the application of a thin adhesive layer from a glue stick. Due to the relative ease of manufacturing and build preparation setup, the advantages of PETG over PP are clear from a manufacturing standpoint.

Selected build parameters are displayed in Table 12. Determined build parameters are relative to a 1.0-mm hot end nozzle and should be modified as such in the case of any significant machine design change. Critical metrics are identified in Table 12. However, more elaborate and complete ".ini" files are included in the Supplementary Documentation.

Table 12. Manufacturing parameters for PETG on a metal-polymer composite Gigabot.

Retraction Parameters Type	Corresponding Slic3r Setting
Print Temperature (°C)	220
Print Speed (mm/s)	40
Layer Height (mm/s)	0.5
Horizontal Shells (Top)	2
Horizontal Shells (Bottom)	3
First Layer Extrusion Width (%)	200
Extrusion Multiplier	×2

Without sufficient accommodation, PETG was noted to string during vector movements and stick to the nozzle. These phenomena caused concern in regards to dimensional stability, printed part accuracy and visual appearance of the printed component. Proper calibration of retention setting and seam locations was required. Table 13 identifies the required print parameter settings to ensure adequate retraction of PETG filament after a vector pass such that no undesired filament was deposited onto the printed part.

Table 13. Manufacturing parameters for PETG on a metal-polymer composite Gigabot.

Parameters Type	Corresponding Slic3r Setting
Length (mm)	10
Lift Z (mm)	0.5
Speed (mm/s)	100
Extra length on restart (mm)	8
Minimum travel after retraction (mm)	0.1
Retract on layer change	Yes
Wipe while retracting	Yes
Seam position	Nearest

4. Discussion

4.1. Practical Application of the Metal-Polymer Composite Gigabot

Attachment of ×5 extruder nozzles to a gantry allows for significant energy/part savings. The developed system contains nearly identical embodied energy and energy consumption when compared to other Cartesian type printer systems on the market (e.g., Lulzbot) [60]. Specifically, comparable systems use a near equivalent amount of NEMA 17 motors: one X-motor, one–two Y-motor(s) and two Z-motors. However, the metal-polymer composite Gigabot allows operators to utilize the embodied energy in the manufacture of multiple components in regards to all X, Y and Z travel movements in all ×5 nozzles simultaneously. Furthermore, the timed-based analysis presented in Section 3.1 displays significant manufacturing time variances between the parameter sets. Most notably are the advantages of utilizing the metal-polymer Gigabot for the manufacture of ×5 components. At 25% in-fill operators printing single components (i.e., one hot end) at a time, 70 min are required for complete manufacture, while 100% requires 125 min for manufacturing. Comparatively, utilizing the full capacity of the metal-polymer Gigabot reduces manufacturing time to 13 min and 24 min for 25% and 100% in-fill, respectively. On a percentage basis, this is a variance of ~438% and ~420% for 25% in-fill and 100% in-fill, respectively. This improved product manufacturing time is an advantage for small lots as could be used in a 3D print shop or part to order factory for small business manufacturing [61]. In addition, this improved embodied energy of manufacturing [62,63] if dispersed would provide an advantage over conventional manufacturing and home-based manufacturing [60,64,65]. At the same time, this methodology points the way toward potential 3D printing-based mass production [66] by ganging many print heads to manufacture identical bespoke products simultaneously [67–70]. This would in theory allow scaling up to the limits of the mechanical strength of the gantry materials to add additional nozzles and the stepper motors to move the assembly of hot ends. This would provide an advantage over smaller producers if the lot size is matched with the number of heads of the 3D printer, while enabling rates approaching more traditional mass manufacturing. However, practically, as the lot sized increases and the geographic market for a particular product expands, the embodied energy of transportation reduces the benefits of reduced embodied energy of manufacturing. Future work is need in environmental life cycle analysis (LCA) to optimize the digital manufacturing mode for energy efficiency and emissions.

4.2. Areas of Improvement and Comparison to Other Technologies

Extruders nozzles mounted on the Y0/Y1 controlled gantry (e.g., the primary cross-head gantry) are fixed upon the x-axis, providing limited mobility relative to one another. Specifically, all five extruders are controlled by the same X_D, $Y0_D$, and $Y1_D$ commands; thus, equivalent movements are required of the head hot end/nozzle. Multi-head FFF systems utilizing Autodesk Project Escher technology [71], for instance the Titan Robotics Cronus 3D Printer [72], allow for hot end individualized positional movements on X, Y and Z for each respective hot end. Current metal-polymer composites designs required a limiting maximum distance of 55 mm in the X-direction. As a result, this limits the maximum part volume printable on the metal-polymer composite Gigabot. To increase the printable part volume, the extents of the printer would have to be enlarged to accommodate hot end linear spacing greater than 55 mm. Extension of the machine mechanical limits would also enable the operator practical utilization of the X_D directional motor at increased hot end spacing. However, build volume optimization processes (i.e. component orientation and 2D build plate layout) can aid operators in the design of manufacturing process parameters within the machine limits. Specifically, the metal-polymer Gigabot retains the ability to print components with their primary (maximum) linear dimension to be oriented perpendicular to the X-direction on the print bed. Effectively, this requires an increased utilization of $Y0_D$ and $Y1_D$ for printing as opposed to X_D. Baumers developed an algorithmic methodology promoting densification of available build plate volume [73]. The methodology employs a selection criterion to promote agglomeration of parts in a build volume [73]. The criterion includes part rotation/orientation, part X/Y positioning coordinates, collision checking and total surface area of the part. In practice, the algorithm selects components to be printed and places them in the proper geometrical coordinates such that their centers of mass are as near as possible to their nearest neighbor [73]. Chernov et al. has developed a practical packing algorithm for classical cutting and packing (C&P) problems. The realized application promotes the minimization of scrap loss during fabrication techniques, such as garment manufacturing, sheet metal cutting and furniture manufacturing. The heuristic algorithms are also applicable to 3D packaging efficiency simulations (i.e., cargo shipments and granular media packaging). For the prescribed models, most are commonly used to analyze simplified polygons fixed in a specific orientation denoted as phi-objects [74]. Similar phi-object models are presented in [75,76]. In FFF printing processes, the operator commonly selects the build orientation based on metrics related to print quality, dimensional stability and mechanical properties. Thus, the slicing software (i.e., Slic3r) is responsible for the X/Y orientation of components to an engineered build plate "density" based on the software algorithms. Thus, while currently developed for non-additive manufacturing processes, Chernov et al.'s methodologies and driving equations could be applied to any metal-polymer composite Gigabot manufacturing system in an effort to optimize build platform part layout under machine constraints. Furthermore, while these methodologies are to be applied to optimize manufacturing processes due to mechanical constraints, in the context of the metal-polymer Gigabot, there are also significant advantages to be discovered from an embodied energy and total capacity utilization (Table 11) standpoint in regards to multi-head ($\times$5) printing.

Bowden sheaths are utilized to provide feed stock material to the five hexagon hot ends. Bowden sheaths reduce the amount of weight on the extruder gantry. A reduction in gantry weight, on any printing system, is generally considered to increase the part quality and positional accuracy as there is less momentum shift between various vector paths. This phenomena is most apparent at faster print speeds. Other composite printers readily available in the marketplace (e.g., Mark Forged) use a direct drive system [77]. At the expense of gantry weight, direct drive printers allow for flexible materials to be extruded. Direct drive accomplishes this by locating an extruder drive motor near the extruder hot end, thus providing sufficient pressure and not allowing flexible material strands (e.g., carbon fiber, fiberglass, high strength high toughness (HSHT) fiberglass and Kevlar to bend and/or flex [77]. The developed metal-polymer composite Gigabot is able to utilize a Bowden system for the feeding of aluminum wire by requiring a pre-engineer tool path and proper fixturing to pull and weave wire through the guide pin into a specified layer geometry. However, the manufactured fixturing bracket for the five hot ends increases the gantry mass greatly, relatively to the delta-style Bowden system [78]. Subsequently, maximum print speeds are not fully realized as the excess mass causes the X_D positioning motor to slip and lose calibration during fast vector changes.

4.3. Future Work

The layer-based manufacturing methodology described is adaptable to other material systems beyond metal/polymer composites. For example, designed reinforcement schemes utilizing carbon fiber and/or fiber glass strands potentially increase the printed composites mechanical properties. A metal-polymer composite Gigabot allows for site-specific placement of reinforcement material for localized strengthening mechanisms. The performance effects of carbon fiber and/or fiber glass embedding require further investigation. Specifically, bonding mechanisms and mechanical property verification (e.g., tensile, yield, elongation and stiffness) are required prior to any implementation in engineering applications.

5. Conclusions

This study described an open-source manufacturing technology that enables the manufacturing of polymer-metal composite components by providing free and open source hardware and software. The developed printing systems achieves metal wire embedment into a polymer matrix 3D printed part via a novel weaving and wrapping method using OpenSCAD and parametric coding for customized g-code commands. The results indicate that utilizing a multi-polymer head system for multi-component manufacturing reduces manufacturing time by ~420–438% and provides dimensionally-uniform components throughout all hot ends/extruders. Maximum dimensional deviation occurs in the x dimension with a value of 0.18 mm on extruder E3. Thus, multi-component manufacturing can produce dimensionally-accurate parts for practical engineering applications.

Supplementary Materials: Supplementary Documentation are available online at www.mdpi.com/2227-7080/5/2/36/s1.

Acknowledgments: The authors would like to thank helpful discussion and technical assistance from S. Snabes, M. Fiedler, B. Wijnen, M. Ohadi and G.C. Anzalone. Financial support was provided by the U.S. Department of Energy Advanced Research Projects Agency-Energy (ARPA-E)/University of Maryland Subaward No. 30353-Z7214003. The views and opinions of the authors herein do not necessarily state or reflect those of the United States

Government or any agency thereof. Additionally, the authors wish to thank James Klausner and Geoffrey Short (both of ARPA-E) for their technical insight and discussions that favorably affected the design of our modules and the general approach to the problem.

Author Contributions: John J. Laureto defined the metal-polymer composite Gigabot's operational protocols to perform experiments and developed samples. Joshua M. Pearce proposed the research idea and analyzed the products. All authors wrote and edited the paper.

Conflicts of Interest: The authors declare no conflicts of interest.

1. Bowyer, A. 3D printing and humanity's first imperfect replicator. *3D Print. Addit. Manuf.* **2014**, *1*, 4–5. [CrossRef]

2. Rundle, G. *Revolution in the Making: 3D Printing, Robots and the Future*; Affirm Press: South Melbourne, Australia, 2014.

3. Jones, R.; Haufe, P.; Sells, E.; Iravani, P.; Olliver, V.; Palmer, C.; Bowyer, A. Reprap—The replicating rapid prototype. *Robotica* **2011**, *29*, 177–191. [CrossRef]

4. Sells, E.; Bailard, S.; Smith, Z.; Bowyer, A.; Olliver, V. RepRap: The replicating rapid prototyper-maximizing customizability by breeding the means of production. In *Handbook of Research in Mass Customization and Personalization*; Pillar, F.T., Tseng, M.M., Eds.; World Scientific: Hackensack, NJ, USA, 2009; Volume 1, pp. 568–580.

5. ASTM F2792-12a. *Standard Terminology for Additive Manufacturing Technologies*; ASTM International: West Conshohocken, PA, USA, 2012.

6. Wittbrodt, B.; Pearce, J. M. The effects of PLA color on material properties of 3D printed components. *Addit. Manuf.* **2015**, *8*, 110–116. [CrossRef]

7. Tymrak, B. M.; Kreiger, M.; Pearce, J. M. Mechanical properties of components fabricated with open-source 3D printers under realistic environmental conditions. *Mater. Des.* **2014**, *58*, 242–246. [CrossRef]

8. Lanzotti, A.; Grasso, M.; Staiano, G.; Martorelli, M. The impact of process parameters on mechanical properties of parts fabricated in PLA with an open-source 3D printer. *Rapid Prototyp. J.* **2015**, *21*, 604–617. [CrossRef]

9. Afrose, M.F.; Masood, S.H.; Iovenitti, P.; Nikzad, M.; Sbarski, I. Effects of part build orientations on fatigue behaviour of FDM-processed PLA material. *Prog. Addit. Manuf.* **2015**, *1*, 1–8. [CrossRef]

10. Tanikella, N.G.; Wittbrodt, B.; Pearce, J.M. Tensile Strength of Commercial Polymer Materials for Fused Filament Fabrication 3D Printing. Unpublished work.

11. Kasture, P.V.; Deole, P.; Irwin, J.L. Case Study Using Open Source Additive Manufacturing (AM) Technology for Improved Part Function. In Proceedings of the ASME 2015 International Mechanical Engineering Congress and Exposition, Houston, TX, USA, 13–19 November 2015.

12. Laureto, J.; Tomasi, J.; King, J.A.; Pearce, J.M. Thermal Properties of 3D Printed Polylactic Acid–Metal Composites. *Prog. Addit. Manuf.* **2017**, 1–15. [CrossRef]

13. Sugavaneswaran, M.; Arumaikkannu, G. Analytical and experimental investigation on elastic modulus of reinforced additive manufactured structure. *Mater. Des.* **2015**, *66*, 29–36. [CrossRef]

14. Duigou, A.L.; Castro, M.; Bevan, R.; Martin, N. 3D printing of wood fibre biocomposites: From mechanical to actuation functionality. *Mater Des.* **2016**, *96*, 106–114. [CrossRef]

15. Taboas, J.M.; Maddox, R.D.; Krebsbach, P.H.; Hollister, S.J. Indirect solid free form fabrication of local and global porous, biomimetic and composite 3D polymer-ceramic scaffolds. *Biomater.* **2003**, *24*, 181–194. [CrossRef]

16. Seitz, H.; Rieder, W.; Irsen, S.; Leukers, B.; Tille, C. Three-dimensional printing of porous ceramic scaffolds for bone tissue engineering. *J. Biomed. Mater. Res.* **2005**, *74*, 782–788. [CrossRef] [PubMed]

17. Habraken, W.J.E.M.; Wolke, J.G.C.; Jansen, J.A. Ceramic composites as matrices and scaffolds for drug delivery in tissue engineering. *Adv. Drug Deliv. Rev.* **2007**, *59*, 234–248. [CrossRef] [PubMed]

18. Liu, X.; Ma, P.X. Polymeric Scaffolds for Bone Tissue Engineering. *Ann. Biomed. Eng.* **2004**, *32*, 477–486. [CrossRef] [PubMed]

19. Da Silva, J.R.C.; da Fonsêca, G.F.G.; de Andrade, M.M. Mechanical tests in thermoplastic elastomers used in 3D printers for the construction of hand prosthesis. In Proceedings of the 2014 Pan American Health Care Exchanges (PAHCE), Brasilia, Brazil, 7–12 April 2014; pp. 1–6.
20. Trachtenberg, J.E.; Mountziaris, P.M.; Miller, J.S.; Wettergreen, M.; Kasper, F.K.; Mikos, A.G. Open-source three-dimensional printing of biodegradable polymer scaffolds for tissue engineering. *J. Biomed. Mater. Res.* **2014**, *102*, 4326–4335. [CrossRef]
21. Wong, J.Y.; Pfahnl, A.C. 3D Printing of Surgical Instruments for Long-Duration Space Missions. *Aviat. Space Environ. Med.* **2014**, *85*, 758–763. [CrossRef] [PubMed]
22. Dimas, L.S.; Bratzel, G.H.; Eylon, I.; Buehler, M.J. Tough Composites Inspired by Mineralized Natural Materials: Computation, 3D printing, and Testing. *Adv. Funct. Mater.* **2013**, *23*, 4629–4638. [CrossRef]
23. Chia, H.N.; Wu, B.M. Recent advances in 3D printing of biomaterials. *J. Biol. Eng.* **2015**, *9*. [CrossRef] [PubMed]
24. Espalin, D.; Muse, D.W.; MacDonald, E.; Wicker, R.B. 3D Printing multifunctionality: structures with electronics. *Int. J. Adv. Manuf. Technol.* **2014**, *72*, 963–978. [CrossRef]
25. Serra, T.; Planell, J.A.; Navarro, M. High-resolution PLA-based composite scaffolds via 3-D printing technology. *Acta Biomaterialia* **2013**, *9*, 5521–5530. [CrossRef] [PubMed]
26. Ma, R.R.; Belter, J.T.; Dollar, A.M. Deposition Manufacturing: Design strategies for multimaterial mechanisms via Three-Dimensional printing and material deposition. *J. Mech. Robot.* **2015**, *7*, 021002.
27. Frazier, W.E. Metal Additive Manufacturing: A Review. *J. Mater. Eng. Perform.* **2014**, *23*, 1917–1928. [CrossRef]
28. Ready, S.; Whiting, G.; Ng, T.N. Multi-Material 3D Printing. *NIP Digit. Fabr. Conf.* **2014**, *2014*, 120–123.
29. Anzalone, G.C.; Zhang, C.; Wijnen, B.; Sanders, P.G.; Pearce, J.M. A Low-Cost Open-Source Metal 3-D Printer. *IEEE Access* **2013**, *1*, 803–810.
30. Haselhuhn, A.S.; Wijnen, B.; Anzalone, G.C.; Sanders, P.G.; Pearce, J.M. In situ formation of substrate release mechanisms for gas metal arc weld metal 3-D printing. *J. Mater. Process. Technol.* **2015**, *226*, 50–59. [CrossRef]
31. Haselhuhn, A.S.; Gooding, E.J.; Glover, A.G.; Anzalone, G.C.; Wijnen, B.; Sanders, P.G.; Pearce, J.M. Substrate Release Mechanisms for Gas Metal Arc Weld 3D Aluminum Metal Printing. *3D Print. Addit. Manuf.* **2014**, *1*, 204–209. [CrossRef]
32. Haselhuhn, A.S.; Buhr, M.W.; Wijnen, B.; Sanders, P.G.; Pearce, J.M. Structure-property relationships of common aluminum weld alloys utilized as feedstock for GMAW-based 3-D metal printing. *Mater. Sci. Eng.* **2016**, *673*, 511–523. [CrossRef]
33. Ding, D.; Pan, Z.; Cuiuri, D.; Li, H. Wire-feed additive manufacturing of metal components: technologies, developments and future interests. *Int. J. Adv. Manuf. Technol.* **2015**, *81*, 465–481. [CrossRef]
34. Wijnen, B.; Anzalone, G.C.; Haselhuhn, A.S.; Sanders, P.G.; Pearce, J.M. Free and open-source control software for 3D motion and processing. *J. Open Res. Softw.* **2015**, *4*. [CrossRef]
35. Laureto, J.; Dessiatoun, S.; Ohadi, M.; Pearce, J. Open Source Laser Polymer Welding System: Design and Characterization of Linear Low-Density Polyethylene Multilayer Welds. *Machines* **2016**, *4*, 14. [CrossRef]
36. Anzalone, G.C.; Wijnen, B.; Pearce, J.M. Multi-material additive and subtractive prosumer digital fabrication with a free and open-source convertible delta RepRap 3-D printer. *Rapid Prototyp. J.* **2015**, *21*, 506–519. [CrossRef]
37. Pearce, J.M. Applications of open source 3D printing on small farms. *Org. Farming* **2015**, *1*, 19–35.
38. Nilsiam, Y.; Haselhuhn, A.; Wijnen, B.; Sanders, P.; Pearce, J. Integrated Voltage—Current Monitoring and Control of Gas Metal Arc Weld Magnetic Ball-Jointed Open Source 3-D Printer. *Machines* **2015**, *3*, 339–351. [CrossRef]
39. Quan, Z.; Wu, A.; Keefe, M.; Qin, X.; Yu, J.; Suhr, J.; Byun, J.-H.; Kim, B.-S.; Chou, T.-W. Additive manufacturing of multi-directional preforms for composites: opportunities and challenges. *Mater. Today* **2015**, *18*, 503–512. [CrossRef]
40. Bayless, J.; Chen, M.; Dai, B. Wire embedding 3D printer. 2010. Available online: http://www.reprap.org/mediawiki/images/2/25/SpoolHead_FinalReport.pdf (accessed on 19 April 2017).
41. Shemelya, C.; Cedillos, F.; Aguilera, E.; Espalin, D.; Muse, D.; Wicker, R.; MacDonald, E. Encapsulated Copper Wire and Copper Mesh Capacitive Sensing for 3-D Printing Applications. *IEEE Sens J.* **2015**, *15*, 1280–1286. [CrossRef]

42. Kim, C.; Espalin, D.; Cuaron, A.; Perez, M.A.; Lee, M.; MacDonald, E.; Wicker, R.B. Cooperative Tool Path Planning for Wire Embedding on Additively Manufactured Curved Surfaces Using Robot Kinematics. *J. Mech. Robot.* **2015**, *7*, 021003–021010. [CrossRef]

43. Saari, M.; Cox, B.; Richer, E.; Krueger, P.S.; Cohen, A.L. Fiber Encapsulation Additive Manufacturing: An Enabling Technology for 3D Printing of Electromechanical Devices and Robotic Components. *3D Print. Addit. Manuf.* **2015**, *2*, 32–39. [CrossRef]

44. Macdonald, E.; Salas, R.; Espalin, D.; Perez, M.; Aguilera, E.; Muse, D.; Wicker, R.B. 3D Printing for the Rapid Prototyping of Structural Electronics. *IEEE Access* **2014**, *2*, 234–242. [CrossRef]

45. Francia, D.; Caligiana, G.; Liverani, A.; Frizziero, L.; Donnici, G. PrinterCAD: A QFD and TRIZ integrated design solution for large size open moulding manufacturing. *Int. J. Interact. Des. Manuf.* **2017**, 1–14. [CrossRef]

46. Re:3D. Available online: http://shop.re3d.org/ (accessed on 19 April 2017).

47. OpenSCAD. Available online: http://www.openscad.org/ (accessed on 19 April 2017).

48. Open Science Framework. Available online: https://osf.io/jvhqt/ (accessed on 19 April 2017).

49. GNU General Public License. Available online: http://www.gnu.org/licenses/gpl-3.0.en.html (accessed on 19 April 2017).

50. MOST Gigabot. Available online: http://www.appropedia.org/Franklin_Firmware_on_GigabotHX:MOST (accessed on 19 April 2017).

51. IC3D Digital Platform. Available online: https://www.ic3dprinters.com/index.html (accessed on 19 January 2017).

52. RAMPS 1.4. Available online: http://reprap.org/wiki/RAMPS_1.4 (accessed on 19 April 2017).

53. Pololu A4988 Stepper Motor Drive Carrier. Available online: https://www.pololu.com/product/1182 (accessed on 19 April 2017).

54. Arduino Mega Board 2560. Available online: https://www.arduino.cc/en/Main/ArduinoBoardMega2560 (accessed on 19 April 2017).

55. KiCad–PcbNew. Available online: http://kicad-pcb.org/discover/pcbnew/ (accessed on 19 April 2017).

56. Slic3r. Available online: http://slic3r.org/ (accessed on 19 April 2017).

57. eSUN–PETG Technical Data Sheet. Available online: http://www.esun3d.net/products/176.html (accessed on 19 April 2017).

58. Gizmo Dorks—PP Technical Data Sheet. Available online: http://gizmodorks.com/polypropylene-3d-printer-filament/ (accessed on 19 April 2017).

59. Cura. Available online: https://ultimaker.com/en/products/cura-software (accessed on 19 April 2017).

60. Petersen, E.; Pearce, J. Emergence of Home Manufacturing in the Developed World: Return on Investment for Open-Source 3-D Printers. *Technologies* **2017**, *5*, 7. [CrossRef]

61. Laplume, A.; Anzalone, G.C.; Pearce, J.M. Open-source, self-replicating 3-D printer factory for small-business manufacturing. *Int. J. Adv. Manuf. Technol.* **2016**, *85*, 633–642. [CrossRef]

62. Kreiger, M.; Pearce, J.M. Environmental Impacts of Distributed Manufacturing from 3-D Printing of Polymer Components and Products. *MRS Proc.* **2013**, *1492*, 85–90. [CrossRef]

63. Megan, K.; Pearce, J.M. Environmental life cycle analysis of distributed three-dimensional printing and conventional manufacturing of polymer products. *ACS Sustain. Chem. Eng.* **2013**, *1*, 1511–1519.

64. Wittbrodt, B.T.; Glover, A.G.; Laureto, J.; Anzalone, G.C.; Oppliger, D.; Irwin, J.L.; Pearce, J.M. Life-cycle economic analysis of distributed manufacturing with open-source 3-D printers. *Mechatron.* **2013**, *23*, 713–726. [CrossRef]

65. Rayna, T.; Striukova, L. From rapid prototyping to home fabrication: How 3D printing is changing business model innovation. *Technol. Forecast Soc. Chang.* **2016**, *102*, 214–224. [CrossRef]

66. Bak, D. Rapid prototyping or rapid production? 3D printing processes move industry towards the latter. *Assembl. Autom.* **2003**, *23*, 340–345. [CrossRef]

67. Hergel, J.; Lefebvre, S. Clean color: Improving multi-filament 3D prints. *Comput. Graphics Forum* **2014**, *33*, 469–478. [CrossRef]

68. Ali, M.H.; Mir-Nasiri, N.; Ko, W.L. Multi-nozzle extrusion system for 3D printer and its control mechanism. *Int. J. Adv. Manuf. Technol.* **2016**, *86*, 999–1010. [CrossRef]

69. Abilgaziyev, A.; Kulzhan, T.; Raissov, N.; Ali, M.H.; Match, W.L.K.; Mir-Nasiri, N. Design and development of multi-nozzle extrusion system for 3D printer. In Proceedings of the 2015 International Conference on Informatics, Electronics & Vision (ICIEV), Fukuoka, Japan, 15–18 June 2015; pp. 1–5.

70. Song, X.; Pan, Y.; Chen, Y. Development of a Low-Cost Parallel Kinematic Machine for Multidirectional Additive Manufacturing. *J. Manuf. Sci. Eng.* **2015**, *137*, 021005–021013. [CrossRef]

71. Autodesk Project Escher. Available online: http://projectescher.com/ (accessed on 19 April 2017).

72. Titan Robotics–Cronus 3D Printer. Available online: http://www.titan3drobotics.com/the-cronus/ (accessed on 19 April 2017).

73. Baumers, M. Economic Aspects of Additive Manufacturing: Benefits, Costs and Energy Consumption. Ph.D. Thesis, Loughborough University, Leicestershire, UK, 2012. Available online: https://dspace.lboro.ac.uk/dspace-jspui/handle/2134/10768 (accessed on 4 April 2017).

74. Chernov, N.; Stoyan, Y.; Romanova, T. Mathematical model and efficient algorithms for object packing problem. *Comput. Geom.* **2010**, *43*, 535–553. [CrossRef]

75. Bennell, J.; Scheithauer, G.; Stoyan, Y.; Romanova, T. Tools of mathematical modeling of arbitrary object packing problems. *Ann. Oper. Res.* **2010**, *179*, 343–368. [CrossRef]

76. Chernov, N.; Stoyan, Y.; Romanova, T.; Pankratov, A. Phi-Functions for 2D Objects Formed by Line Segments and Circular Arcs. *Adv. Oper. Res.* **2012**, *2012*, 26. [CrossRef]

77. Markforged Composite 3D Printer. Available online: https://markforged.com/why-markforged/ (accessed on 19 April 2017).

78. MOST Athena Delta—End Effector. Available online: http://www.appropedia.org/Athena_Effector_Assembly (accessed on 19 April 2017).

technologies

MDPI

Article

A Robust Multifunctional Sandwich Panel Design with Trabecular Structures by the Use of Additive Manufacturing Technology for a New De-Icing System

Carlo Giovanni Ferro [1,*], Sara Varetti [2,3], Fabio Vitti [1], Paolo Maggiore [1], Mariangela Lombardi [3], Sara Biamino [3], Diego Manfredi [4] and Flaviana Calignano [4]

[1]   Department of Mechanical Engineering and Aerospace (DIMEAS), Politecnico di Torino,
      Corso Duca Degli Abruzzi 24, 10129 Turin, Italy; fabio.vitti.sc@gmail.com (F.V.); paolo.maggiore@polito.it (P.M.)
[2]   3D-New Technologies (3D-NT), via Livorno 60, 10144 Turin, Italy; sara.varetti@polito.it
[3]   Department of Applied Science and Technology (DISAT), Politecnico di Torino,
      Corso Duca Degli Abruzzi 24, 10129 Turin, Italy; mariangela.lombardi@polito.it (M.L.);
      sara.biamino@polito.it (S.B.)
[4]   Center for Sustainable Future Technologies CSFT@Polito, Istituto Italiano di Tecnologia (IIT),
      Corso Trento 21, 10129 Turin, Italy; Diego.Manfredi@iit.it (D.M.); Flaviana.Calignano@iit.it (F.C.)
*   Correspondence: carlo.ferro@polito.it; Tel.: +39-11-090-6858

Academic Editor: Manoj Gupta
Received: 29 January 2017; Accepted: 7 June 2017; Published: 15 June 2017

Abstract: Anti-ice systems assure a vital on-board function in most aircraft: ice prevention or de-icing is mandatory for all aerodynamic surfaces to preserve their performance, and for all the movable surfaces to allow the proper control of the plane. In this work, a novel multi-functional panel concept which integrates anti-icing directly inside the primary structure is presented. In fact, constructing the core of the sandwich with trabecular non-stochastic cells allows the presence of a heat exchanger directly inside the structure with a savings in weight and an improvement in thermal efficiency. This solution can be realized easily in a single-piece component using Additive Manufacturing (AM) technology without the need for joints, gluing, or welding. The objective of this study is to preliminarily investigate the mechanical properties of the core constructed with Selective Laser Melting (SLM); through the Design of Experiment (DOE), different design parameters were varied to understand how they affect the compression behaviour.

Keywords: Additive Manufacturing (AM); Selective Laser Melting (SLM); advanced structures; anti-ice systems; lattice structures; Design of Experiment

1. Introduction

Icing on aircraft is a concrete and severe problem and has led to direct or indirect hazards and fatal crashes during recent years [1,2]. The phenomenon is caused by supercooled droplets of water normally present in clouds that strike the leading edges of aircraft and freeze on impact [3].

The ice type formed on aircraft parts depends on the weather and the temperature of the contact surface [4–8]. Among the most common ice types are clear ice, formed from big water droplets, most tenacious and heavy to remove and SLD (Supercooled Large Droplets) ice, similar to clear ice but more extended, and can cause severe damage as in the crash of the American Eagle Flight 4184 [9]. Another common ice type is rime ice, which is formed by smaller diameter droplets and is easy to remove due to its fragility. Finally, mixed ice, with intermediate characteristics, is also common.

One of the main risks due to icing during flight is aerodynamic hazard, beacause the formation of ice may cause the modification of the airfoil profile and so increases the drag united with the decrease in lift. Another possible risk factor is represented by structural systems hazards, wich are related to the additional weight that unbalances structural parts affected by ice formation (rotor propellers, antennas, main rotor of helicopters, etc.). Systems hazards are a further source of risk, and occur when ice causes the lock of surfaces that may lead to the loss of control of the flight.

Anti-icing systems need to protect the exposed parts from ice and rain, namely wing leading edges, horizontal and vertical stabilizers, engine nacelle leading edges (engine fans have to face the FOD (Foreign Object Debris) problem), propellers, air data probe (Pitot), windows of the flight deck, antennas and water and waste systems. The most frequently used anti-icing systems are thermal; in fact, the use of heating fluids or electrical resistances for maintaining the adequate temperature on the outer surface of the wing is common to many patented systems. According to civil regulation [10], the system must withstand ice formation in an atmosphere at $-9.4\ ^\circ$C, with a LWC (Liquid Water Content) of $0.5\ \mathrm{g/m^3}$ and a droplet mean volumetric diameter (DMVD) of 20 microns. A wide range of patents describe the use of hot air bled from the compressor of the turbo-engine as heat carrier. The hot air enters structures necessary for the transport of heat on the outer surface. These structures may be inflatable elements affixed on the leading edge insufflated with hot air [11], or porous leading edges through which the anti-ice fluid flows [12]. Other structures are pipes that carry hot air up to the inner surface of the panel [13], leading edges with internal pipes that favour the forced air path next to the outer sheet [14], and finally a series of systems which require the use of ducts and interstices used as de-icing installations [15].

In this work, a new patented [16] system for the de-icing and the anti-icing of airplanes is presented, together with its method of fabrication. The novelty of this system is the integration of hot air passageways and feeding tubes in a single-piece structural panel of the wing leading edge, without the necessity of welding or other kinds of joining. This novel system allows an important savings in the mass of the primary structure, and an improvement of the thermal efficiency of the de-icing function. The system is a sandwich panel composed by core, outer, and inner skin. The core is produced with a non-stochastic lattice structure, while the external skin is the aerodynamic surface, and the internal skin integrates the feeding tubes that collect the hot air bled from the compressor.

The use of sandwich panels with trabecular structure as a core ensures a high specific surface area that optimizes heat exchange. The panel must be manufactured in a single material and in one piece. The realization of structures of this type, with a controlled porosity, reduced dimensions of the details, and articulated geometry, is very difficult to achieve with traditional foundry and molding techniques. These needs may instead be exhaustively fulfilled by additive manufacturing technologies, in particular selective laser melting (SLM), a laser powder bed fusion (LPBF) process that allows the manufacturing of metal components through selective melting of powders, layer by layer [17,18]. Fusion occurs only in the areas necessary for the realization of the component according to the information obtained by the stereo lithography (STL) digital format of the 3D model. The energy required to melt the powder is supplied by a laser beam [19,20] in inert atmosphere (N_2, Ar). Additive manufacturing is the only technology which allows the realization of a single-piece panel in a single material and with a complex geometry core.

In the literature, there are several works that describe the feasibility and the behaviour of periodic lattice structures using different materials and additive technologies such as SLM and electron beam melting (EBM). The studied geometries range from BCC (body-centered cubic) and FCC (face-centered cubic), with their variants, to more complex cells such as gyroid and diamond [21,22]. Many papers report the results of FE models and mechanical tests, especially compression tests, on samples with lattice structures, through the study of the fracture mechanics [23–30].

SLM technology was used in the study of trabecular structures of 316L stainless steel [31–37] and aluminum alloys such as AlSi10Mg [21,22,38–40]. Al alloys are particularly suited for the application described in the present paper, as Al alloys show excellent thermal conductivity and excellent corrosion resistance. The purpose of the investigations is to choose the best core structure that constitutes the sandwich panel, so different types of specimens were produced with varying strut thickness, cell type, and cell size. An analysis of their compressive behaviour was conducted to assess which of these factors is most influential on mechanical properties.

2. Material and Methods

In order to evaluate the properties of the trabecular core, mechanical tests were carried out on different cell types. Taking as input both technical constraint and thermal-mechanical request, a 2^3 Full Factorial DOE (Design Of Experiment) was imposed. Two promising cell types were chosen to evaluate the compressive behaviour varying strut size (1 mm and 1.2 mm) and cell size (4 mm and 5 mm). This approach was applied to produce lattice samples and to experimentally test the mechanical properties. In Figure 1, two of the 3D models of the specimens produced are presented. The geometries of the models correspond to a body-centered cubic (bcc) structure and to a body-centered cubic with vertical struts along the Z axis (bcc-z). Eight kinds of specimens were realized, with two samples for each type. Table 1 lists the specimens with the respective characteristics (cell type, cell size, and strut size).

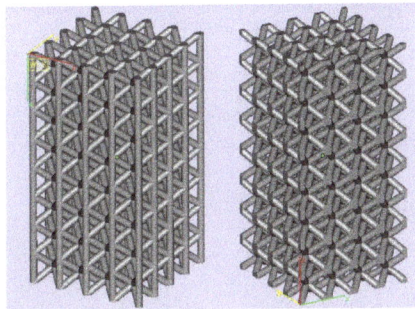

Figure 1. 3D models of trabecular specimens with bcc-z and bcc cell type.

Table 1. List of specimens produced and their characteristics.

Samples Name	Cell Size (mm)	Strut Size (mm)	Cell Type
4-1-bcc-z	4	1	bcc-z *
4-1.2-bcc-z	4	1.2	bcc-z *
5-1-bcc-z	5	1	bcc-z *
5-1.2-bcc-z	5	1.2	bcc-z *
4-1-bcc	4	1	bcc **
4-1.2-bcc	4	1.2	bcc **
5-1-bcc	5	1	bcc **
5-1.2-bcc	5	1.2	bcc **

* Body-centered cubic with vertical struts along Z axis; ** Body-centered cubic.

All samples were made with the EOS machine M270 Dual Mode Version with SLM technology. They grew in height, with a square base resting on the plate. The thickness of the powder bed layer was 30 μm. The powder used was AlSi10Mg, a typical casting alloy that offers good strength, hardness, and is also used for parts subjected to high loads. AlSi10Mg is suitable for applications that require both good thermal properties and low weight. The dimensions of the specimens constructed were 20 mm of base size and 40 mm of height. All compression tests were performed with a Zwick Roell machine: the samples were compressed along the Z axis, with the XY faces in contact with the plates. A preload of 1 kN was imposed, and then a load cell of 50 kN was applied by setting a constant displacement of 1 mm/min.

From mechanical characteristics obtained for each specimen, the influence of all factors on the performance can be estimated. The main effect was calculated for the specific elastic modulus (E/ρ), specific maximum stress (σ_{max}/ρ), and specific stress corresponding to a permanent deformation of

0.2% ($\sigma_{0.2}/\rho$). The plots, realized with Minitab, show the effect of strut size, cell size, and cell type on specific mechanical properties of the specimens. The influence of each factor on the final parameter was obtained from the slope of the line in the graph (in modulus): the greater the slope, the greater the influence. The slope is calculated as $\Delta Y/\Delta X$, where ΔY is the difference between the average values of the property (example: $(E/\rho)_{1.2,average} - (E/\rho)_{1,average} = \Delta Y$), and ΔX is assumed unitary in order to compare the slopes. The same procedure was used for all data and all factors.

Further analysis of the effects of each factor was conducted with Pareto charts. The Pareto principle says that most of the effect is due to a small number of causes: this means that 80% of the effects are achieved by using 20% of the factors. The use of this graph allows the evaluation of not only the influence of factors, but also the influence of their interactions. In fact, the effect of factors and interactions is not the same; some have greater influence and others minor.

3. Results

The obtained results led to qualitative inferences useful for the design of new experiments on the cellular core. The core is the most important structural part, and its compressive strength, which is the subject of the first part of characterization, is fundamental. Figure 2 illustrates the entire panel with the internal lattices, the external skins, and the feeding tubes integrated into a single component. In the structure there are no rivets and joinings which are present in the traditional thermal systems (e.g., that of Piaggio P180 shown in the Figure 3), and so unpleasant phenomena like cracks or damages can be overcome.

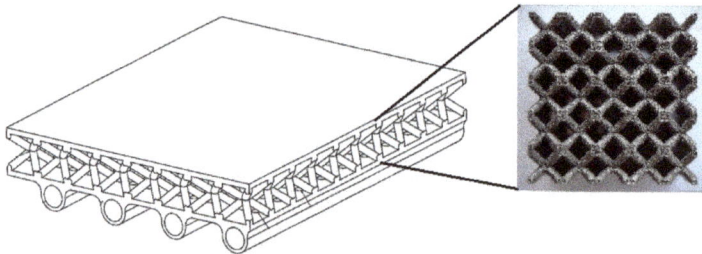

Figure 2. Section of the panel.

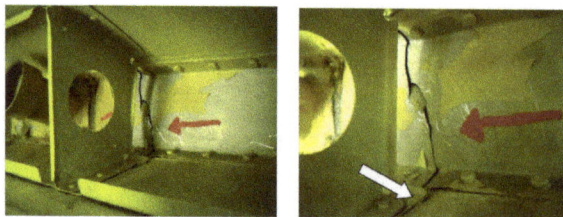

Figure 3. Internal structure of Piaggio thermal anti-icing system, with cracks evidenced by red arrows, courtesy of [41].

In Table 2, pure and specific mechanical properties are reported. The specific values were obtained by dividing E, σ_{max}, $\sigma_{0.2}$ for the density of each specimen. The best elastic modulus and the greatest stresses (σ_{max} and $\sigma_{0.2}$) were obtained for the specimen with 4 mm cell size, 1.2 mm strut size, and with bcc-z cell geometry. The DOE approach allows an easy evaluation of the single effect of each variable, in order to identify the most influential factor.

Table 2. Mechanical properties of the specimens tested.

Samples	E (MPa)	σ_{max} (MPa)	$\sigma_{0.2}$ (MPa)	E/ρ (MPa · dm³/kg)	σ_{max}/ρ (MPa · dm³/kg)	$\sigma_{0.2}/\rho$ (MPa · dm³/kg)
4-1-bcc-z	1054	29	20	1576	43	30
4-1.2-bcc-z	1465	54	37	1587	59	40
5-1-bcc-z	565	15	13	1263	33	28
5-1.2-bcc-z	933	29	22	1463	45	35
4-1-bcc	216	11	10	357	18	16
4-1.2-bcc	717	24	18	852	29	22
5-1-bcc	63	4	4	154	11	11
5-1.2-bcc	253	10	9	434	17	15

The main effect charts in Figure 4a,c,e show the effect of strut size, cell size, and cell type on E/ρ, σ_{max}/ρ, and $\sigma_{0.2}/\rho$ for the tested specimens. The slopes of the lines indicate that the increase in the strut size has a positive effect on the mechanical behaviour, leading to a more dense and stronger trabecular structure. For the same reason, the increase of the cell size leads to a reduction in the density of the foam, with a decrease in elastic modulus, in σ_{max}, and in $\sigma_{0.2}$. Finally, a positive result can be noticed looking at the effect of cell type: the presence of vertical struts leads to a stiffer foam. Comparing the slopes, it appears that the factor that most influences the specific properties is the cell type. Passing from a bcc cell type to a bcc-z cell type, an increase in E/ρ, σ_{max}/ρ, and $\sigma_{0.2}/\rho$ was noticed. This increase is greater than that obtained, for example, by varying the strut size from 1 to 1.2 mm. The influence of cell size and of strut size, instead, is similar. The slopes reported for E/ρ, σ_{max}/ρ, and $\sigma_{0.2}/\rho$ follow the same behaviour: a stiffer and more dense structure leads to an improvement of the mechanical properties. In all cases, the factor that most influences the parameters is the cell type.

The dot-plots reported in Figure 4b,d,f show a low dispersion of the data for E/ρ, σ_{max}/ρ, and $\sigma_{0.2}/\rho$, with the same cell shape, while a high dispersion is presented for the other two factors. This is also due to the decisive role of the cell shape on mechanical behavior, compared to the effect of strut size and cell size.

Further analysis of the specimens was conducted, plotting the Pareto charts, reported in Figure 5. Regarding the specific elastic modulus, the greatest influential factor is the cell type, and it is also the only one above the significance value. Immediately after single factors, there is a second-order interaction between cell type and strut size, and then there are the third-order interaction and all the other second-order interactions. The Pareto charts for σ_{max}/ρ and $\sigma_{0.2}/\rho$ (Figure 5) show that the greatest influence on the output is given by the cell type, but different values are reported: for σ_{max}/ρ, all the single factors seem to be significant, while for $\sigma_{0.2}/\rho$, only the cell type seems to be. Moreover, the second factor that influences the specific stresses is the strut size, but not the cell size, as seen before for the specific elastic modulus.

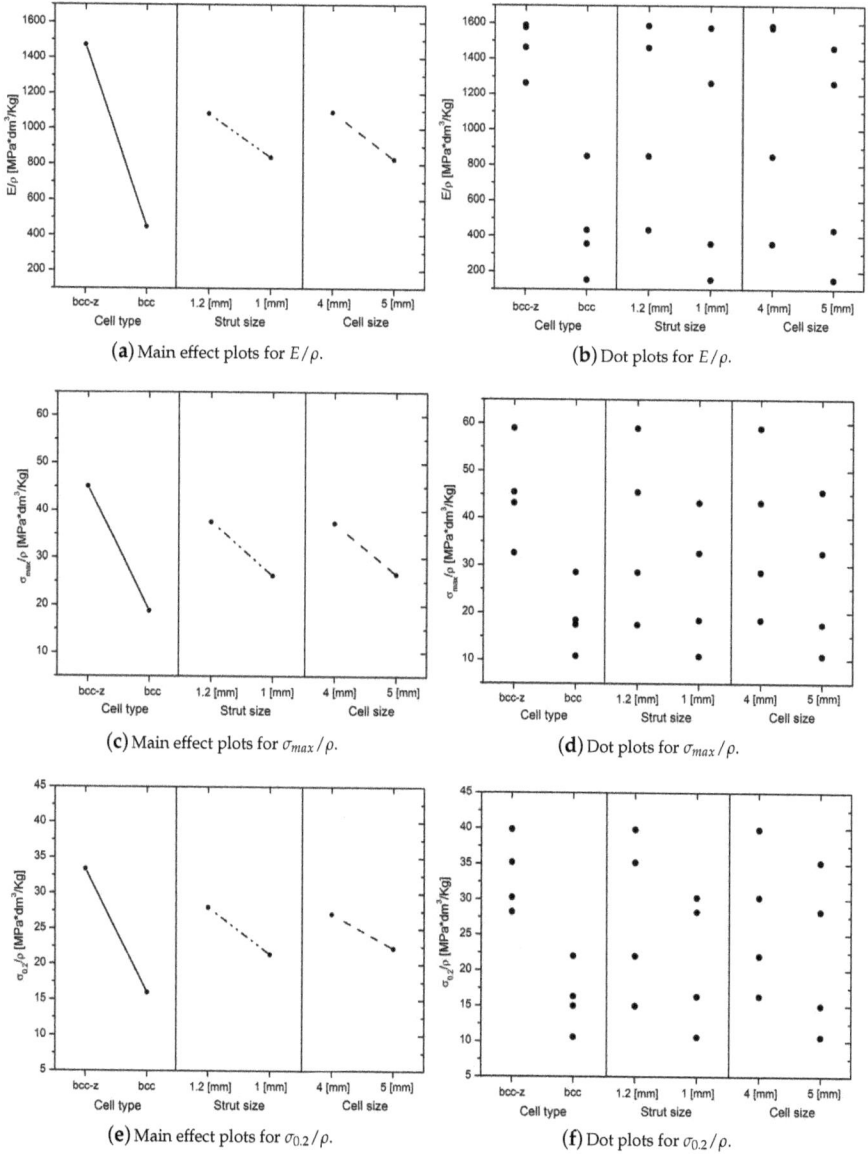

Figure 4. Main effect plots and dot-plots for the specific elastic modulus, specific maximum stress, and specific stress for 0.2% of deformation.

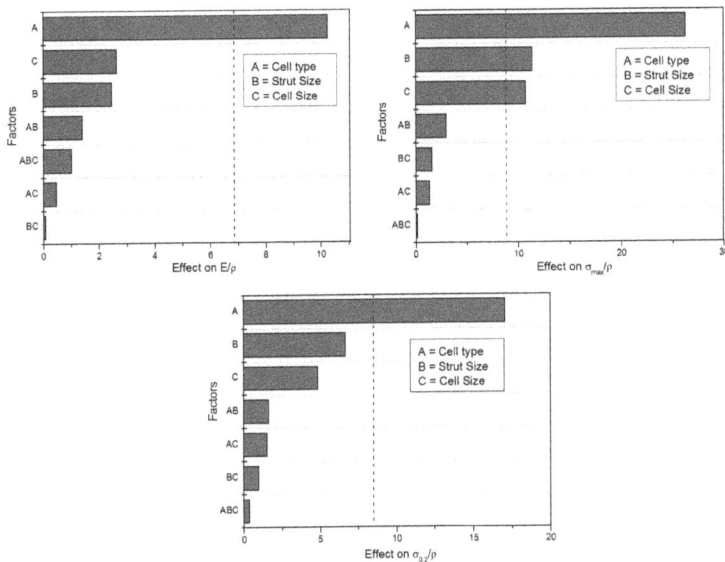

Figure 5. Pareto charts for E/ρ, $\sigma_{0.2}/\rho$, and σ_{max}/ρ.

4. Discussion

From the results reported in the previous section, few assumptions can be summarized, providing general rules. The first consideration is that the presence of vertical struts, although they have a buckling failure, leads to the best mechanical performance, both in terms of specific elastic modulus and in terms of strengths. The second observation is closely linked to the first one: the most important design variable is the cell type and so the struts orientation, with a magnitude superior to the other two factors. The third point is that a more dense structure (thicker struts and smaller cells) always gives better results. This fact is pretty understandable, since these changes imply an increase in the effective resistant area and a rise in the global amount of density. With these considerations, it is possible to correctly design a core for the sandwich panel, allowing it to sustain the normal pressure of the aerodynamic load during flight, reducing masses, and improving mechanical performances.

5. Conclusions and Further Improvements

In this paper, the idea of a multi-functional panel is described together with an experimental evaluation of some design parameters by compression tests. The additive manufacturing technology is tested for the production of non-stochastic cells, and it appears suitable to the purpose. The experiments aimed at the characterization of the panels, starting with the present work, will continue with the construction of other specimens and with bending, tension, and impact tests. Moreover, future works will include a multidisciplinary optimization of the trabeculae, considering thermal phenomena (with lumped parameters models and CFD (Computational Fluid Dynamics) analysis) and mechanical behaviour of the panel. At the end, a de-icing test will be carried out by subjecting the sandwich panel with a trabecular core to humidity, temperature, and pressure conditions identical to those in flight condition. The experiments will allow the de-icing effectiveness to be analysed.

Author Contributions: Carlo Giovanni Ferro, Sara Varetti and Fabio Vitti conceived and designed the experiments under the supervision of Paolo Maggiore, Mariangela Lombardi and Sara Biamino. Sara Varetti performed the experiments and analyzed the data with Carlo Giovanni Ferro. Flaviana Calignano and Diego Manfredi contributed with materials and specimens. All authors discussed and analyzed the results, contributed to writing and approved the final manuscript.

Conflicts of Interest: The authors declare no conflict of interest.

1. Bureau d'Enquêtes et d'Analyses pour la sécurité de l'aviation civile (BEA). Final Report on the Accident on 1 June 2009 to the Airbus A330-203 Registered F-GZCP Operated by Air France Flight AF 447 Rio de Janeiro-Paris. BEA 2012, 10. Available online: https://www.bea.aero/en/investigation-reports/notified-events/detail/event/accident-of-an-airbus-a330-203-registered-f-gzcp-and-operated-by-air-france-crashed-into-the-atlanti/ (accessed on 14 June 2017).
2. Transportation Safety Board of Canada (TSBC). Final Report on the Transportation Safety Board of Canada's Investigation into the Accident Involving an Air Canada Flight 646 Regional Jet at Fredericton, New Brunswick, on the Night of 16 December 1997. TSBC 1999, 10. Available online: http://www.tsb.gc.ca/ENG/medias-media/communiques/aviation/1999/comm_a97h0011.asp (accessed on 14 June 2017).
3. Federal Aviation Administration (FAA). In *Aircraft Icing Handbook*; FAA Technical Center: Atlantic City, NJ, USA, 1991.
4. Politovich, M.K. Aviation Meteorology: Aircraft Icing. In *Encyclopedia of Atmospheric Sciences*; National Center for Atmospheric Research, Boulder: Boulder, CO, USA, 2015; pp. 160–165.
5. Cao, Y.; Wu, Z.; Su, Y.; Xu, Z. Aircraft flight characteristics in icing conditions. *Progress Aerosp. Sci.* **2015**, *74*, 62–80.
6. Li, X.; Bai, J.; Hua, J.; Wang, K.; Zhang, Y. A spongy icing model for aircraft icing. *Chin. J. Aeronaut.* **2014**, *27*, 40–51.
7. Srensen, K.L.; Blanke, M.; Johansen, T.A. Diagnosis of Wing Icing Through Lift and Drag Coefficient Change Detection for Small Unmanned Aircraft. *IFAC-PapersOnLine* **2015**, *48*, 541–546.
8. Parent, O.; Ilinca, A. Anti-icing and de-icing techniques for wind turbines: Critical review. *Cold Reg. Sci. Technol.* **2011**, *65*, 88–96.
9. National Transport and Safety Authority (NTSA). National Transportation Safety Board (NTSB) Aircraft Accident Report, Monday 31 October 1994. Transportation Safety Board of Canada (TSBC) 1996, 10. Available online: https://www.ntsb.gov/investigations/AccidentReports/Pages/AAR9601.aspx (accessed on 14 June 2017).
10. Federal Aviation Administration (FAA). *MIL-A-9482 Anti-Icing Equipment for Aircraft, Heated Surface Type, General Specification*; FAA Technical Center: Atlantic City, NJ, USA, 1981.
11. Goodman, J. Means for Preventing Ice Formation on Aircraft Wings. U.S. Patent 2,328,079, 31 August 1943.
12. Garrison, M.E. Airplane Wing Deicing Means. U.S. Patent 2,390,093, 4 December 1945.
13. Schmidt, H.F. Airplane Anti-Icing System. U.S. Patent 2,447,095, 17 August 1948.
14. Ayers, M.; Barrick, R.E. Airplane Wing Leading Edge Costruction. U.S. Patent 2,470,128, 17 May 1949.
15. Dodson, P.A.; Smith, L.H. Deicing Wing Costruction. U.S. Patent 2,478,878, 9 August 1949.
16. Giovanni, F.C.; Varetti, S.; Vitti, F.; Maggiore, P. An Aircraft Equipped with Astructurally Integrated De-Icing System. IT102016000098196, 2016.
17. Ford, S.; Despeisse, M. Additive manufacturing and sustainability: An exploratory study of the advantages and challenges. *J. Clean. Prod.* **2016**, *137*, 1573–1587.
18. Moylan, S.; Whitenton, E.; Lane, B.; Slotwinski, J. Infrared Thermography for Laser-Based Powder Bed Fusion Additive Manufacturing Processes. In Proceedings of the 40th Annual Review of Progress in Quantitative Nondestructive Evaluation, Baltimore, MD, USA, 21–26 July 2013; pp. 1191–1196.
19. Kempen, K.; Thijs, J.; Van Humbeeck, J.; Kruth, J.-P. Mechanical properties of AlSi10Mg produced by Selective Laser Melting. *Phys. Proced.* **2012**, *39*, 439–446.
20. Wong, K.V.; Hernandez, A. A Review of Additive Manufacturing. *ISRN Mech. Eng.* **2012**, *2012*, 208760, doi:10.5402/2012/208760.
21. Yan, C.; Hao, L.; Hussein, A.; Young, P.; Huang, J.; Zhu, W. Microstructure and mechanical properties of aluminium alloy cellular lattice structures manufactured by direct metal laser sintering. *Mater. Sci. Eng. A* **2015**, *628*, 238–246.
22. Yan, C.; Hao, L.; Hussein, A.; Bubb, S.L.; Young, P.; Raymont, D. Evaluation of light weight AlSi10Mg periodic cellular lattice structures fabricated via direct metal laser sintering. *J. Mater. Process. Technol.* **2014**, *214*, 856–864.

23. Cansizoglu, O.; Harrysson, O.; Cormier, D.; West, H.; Mahale, T. Properties of Ti6Al4V non stochastic lattice structures fabricated via electron beam melting. *Mater. Sci. Eng. A* **2008**, *492*, 468–474.

24. Murr, L.E.; Gaytan, S.M.; Medina, F.; Martinez, E.; Martinez, J.L.; Hernandez, D.H.; Machado, B.I.; Ramirez, D.A.; Wicker, R.B. Characterization of Ti6Al4V open cellular foams fabricated by additive manufacturing using electron beam melting. *Mater. Sci. Eng. A* **2010**, *527*, 1861–1868.

25. Zhao, S.; Li, S.J.; Hou, W.T.; Hao, Y.L.; Yang, R.; Misra, R.D.K. The influence of cell morphology on the compressive fatigue behavior of Ti6Al4V meshes fabricated by electron beam melting. *J. Mech. Behav. Biomed. Mater.* **2016**, *59*, 251–264.

26. Li, S.J.; Murr, L.E.; Cheng, X.Y.; Zhang, Z.B.; Hao, Y.L.; Yang, R.; Medina, F.; Wicker, R.B. Compression fatigue behavior of Ti–6Al–4V mesh arrays fabricated by electron beam melting. *Acta Mater.* **2012**, *60*, 793–802.

27. Gorny, B.; Niendorf, T.; Lackmann, J.; Thoene, M.; Troester, T.; Maier, H.J. In situ characterization of the deformation and failure behavior of non-stochastic porous structures processed by selective laser melting. *Mater. Sci. Eng. A* **2011**, *528*, 7962–7967.

28. Lhuissier, P.; Formanoir, C.; Martin, G.; Dendievel, R.; Godet, S. Geometrical control of lattice structures produced by EBM through chemical etching: Investigations at the scale of individual struts. *Mater. Des.* **2016**, *110*, 485–493.

29. Brenne, F.; Niendorf, T.; Maier, H.J. Additively manufactured cellular structures: Impact of microstructure and local strains on the monotonic and cyclic behavior under uniaxial and bending load. *J. Mater. Process. Technol.* **2013**, *213*, 1558–1564.

30. Suard, M.; Martin, G.; Lhuissier, P.; Dendievel, R.; Vignat, F.; Blandin, J.-J.; Villeneuve, F. Mechanical equivalent diameter of single struts for the stiffness prediction of lattice structures produced by Electron Beam Melting. *Addit. Manuf.* **2015**, *8*, 124–131.

31. Smith, M.; Guan, Z.; Cantwell, W.J. Finite element modelling of the compressive response of lattice structures manufactured using the selective laser melting technique. *Int. J. Mech. Sci.* **2013**, *67*, 28–41.

32. Yan, C.; Hao, L.; Hussein, A.; Young, P.; Raymont, D. Advanced lightweight 316L stainless steel cellular lattice structures fabricated via selective laser melting. *Mater. Des. J.* **2014**, *55*, 533–541.

33. Yan, C.; Hao, L.; Hussein, A.; Raymont, D. Evaluations of cellular lattice structures manufactured using selective laser melting. *Int. J. Mach. Tools Manuf.* **2012**, *62*, 32–38.

34. Alsalla, H.; Hao, L.; Smith, C. Fracture toughness and tensile strength of 316 L stainless steel cellular lattice structures manufactured using the selective laser melting technique. *Mater. Sci. Eng. A* **2016**, *669*, 1–6.

35. Gumruk, R.; Mines, R.A.W. Compressive behaviour of stainless steel micro-lattice structures. *Int. J. Mech. Sci.* **2013**, *68*, 125–139.

36. Gümrük, R.; Mines, R.A.W.; Karadeniz, S. Static mechanical behaviours of stainless steel micro-lattice structures under different loading conditions. *Mater. Sci. Eng. A* **2013**, *586*, 392–406.

37. McKown, S.; Shen, Y.; Brookes, W.K.; Sutcliffe, C.J.; Cantwell, W.J.; Langdon, G.S.; Nurick, G.N.; Theobald, M.D. The quasi-static and blast loading response of lattice structures. *Int. J. Impact Eng.* **2008**, *35*, 795–810.

38. Maskery, I.; Aboulkhair, N.T.; Aremu, A.O.; Tuck, C.J.; Ashcroft, I.A.; Wildman, R.D.; Hague, R.J.M. A mechanical property evaluation of graded density Al-Si10-Mg lattice structures manufactured by selective laser melting. *Mater. Sci. Eng. A* **2016**, *670*, 264–274.

39. Qiu, C.; Yue, S.; Adkins, N.J.E.; Ward, M.; Hassanin, H.; Lee, P.D.; Withers, P.J.; Attallah, M.M. Influence of processing conditions on strut structure and compressive properties of cellular lattice structures fabricated by selective laser melting. *Mater. Sci. Eng. A* **2015**, *628*, 188–197.

40. Leary, M.; Mazur, M.; Elambasseril, J.; McMillan, M.; Chirent, T.; Sun, Y.; Qian, M.; Easton, M.; Brandt, M. Selective laser melting (SLM) of AlSi12Mg lattice structures. *Mater. Des.* **2016**, *98*, 344–357.

41. Vacca, A. P.180 Main Wing Anti-Ice System: Analysis and Improvements. Master's Thesis, University of Genoa, Genova, Italy, 2013.

technologies

[MDPI]

Article

Monitoring Approach to Evaluate the Performances of a New Deposition Nozzle Solution for DED Systems

Federico Mazzucato [1,*], Simona Tusacciu [2], Manuel Lai [2], Sara Biamino [3], Mariangela Lombardi [3] and Anna Valente [1]

1   SUPSI, ISTePS-Institute of Systems and Technologies for the Sustainable Production, Galleria 2, Manno (6928), Switzerland; anna.valente@supsi.ch
2   IRIS S.r.l., Via Papa Giovanni Paolo II, 26, 10043 Orbassano, TO, Italy; simona.tusacciu@irissrl.org (S.T.); manuel.lai@irissrl.org (M.L.)
3   Department of Applied Science and Technology, Politecnico di Torino, Corso Duca degli Abruzzi, 24, 10129 Torino, Italy; sara.biamino@polito.it (S.B.); mariangela.lombardi@polito.it (M.L.)
*   Correspondence: federico.mazzucato@supsi.ch; Tel.: +41-(0)58-666-67-08

Academic Editors: Salvatore Brischetto, Paolo Maggiore and Carlo Giovanni Ferro
Received: 18 April 2017; Accepted: 25 May 2017; Published: 31 May 2017

Abstract: In order to improve the process efficiency of a direct energy deposition (DED) system, closed loop control systems can be considered for monitoring the deposition and melting processes and adjusting the process parameters in real-time. In this paper, the monitoring of a new deposition nozzle solution for DED systems is approached through a simulation-experimental comparison. The shape of the powder flow at the exit of the nozzle outlet and the spread of the powder particles on the deposition plane are analyzed through 2D images of the powder flow obtained by monitoring the powder depositions with a high-speed camera. These experimental results are then compared with data obtained through a Computational Fluid Dynamics model. Preliminary tests are carried out by varying powder, carrier, and shielding mass flow, demonstrating that the last parameter has a significant influence on the powder distribution and powder flow geometry.

Keywords: direct deposition machine; scanner systems; metallic additive manufacturing

1. Introduction

In recent years, additive manufacturing (AM) technologies for the production of metal parts have drawn an enormous surge of industrial interest; however, process reliability and component quality are not enough for mass production [1]. In opposition to systems able to selectively melt a powder bed, high build rates and larger part volumes can be obtained by exploiting direct energy deposition (DED) technologies [2,3]. In this technology, previously trademarked as laser engineered net shaping (LENS), laser metal deposition (LMD), or direct metal deposition (DMD), a heat source generates a melt pool on a metallic surface into which feedstock is deposited with the consequent building of the parts layer by layer. At the moment, several solutions are available, using wire or powder as feedstock; in former versions, the processes could be considered a type of welding technology, whereas the latter ones are very similar to laser cladding technologies.

In view of the introduction of AM machines to the industrial world, it is necessary to study and develop monitoring and control systems capable of guaranteeing series production capability and reproducibility of these innovative processes. For this reason, recent AM technology reviews have repeatedly called for real-time, closed loop process controls and sensors to ensure quality, consistency, and reproducibility across AM machines [4].

DED systems employing powder feeding are highly sensitive to working conditions; any variation of process parameters during the deposition and melting of the metal powders can influence the quality of the component, compromising the process efficiency [5]. In particular, the deposition quality can be affected by intrinsic parameters (related to the properties of the substrate and the metallic powder such as geometry, thermal diffusivity, absorptivity, thermal conductivity, and heat capacity) and extrinsic parameters (related to the laser, the powder feeder, and the positioning system). In particular, extrinsic parameters are strictly related to the specific print head of the employed AM machine, in which powder injection nozzles and the laser beam are assembled according different schemes [6]. Indeed, common DED or cladding powder injection systems can present:

- an off-axis configuration, in which a single powder flow laterally passes through the laser beam;
- a continuous coaxial configuration, in which a conical powder flow surrounds and interacts with the laser beam;
- a discontinuous coaxial configuration, with several powder flows from different injection nozzles (i.e., multiple nozzle deposition) distributed around the laser beam (up to eight in a robotized laser-based direct metal deposition system recently developed [7]).

The off-axis powder injection implies a strong relation between the deposition rate and the scan direction [8]; in the continuous coaxial version, as opposed to the discontinuous axial one, the tilting of the powder injection nozzle is restricted, limiting the potentialities of DED [6]. Multiple nozzle deposition is suitable for parts with high geometrical complexity, even if it is more difficult to ensure flow uniformity and direct the powder to a specific region of interest [9]. In the case of both continuous coaxial configuration and discontinuous coaxial configuration, the spot of the powder flow in correspondence with the deposition plane usually ranges between 5 mm and 8 mm [6,10–13].

As already stated, the process efficiency can be influenced by the design of the powder injection nozzle, since it is capable of modifying interactions among the metallic particles, laser beam, and melt pool [14]. Further, the control of extrinsic parameters can be useful for adjusting deposition errors during the process. For this reason, in order to improve the process reliability and quality, closed loop control systems should be introduced in DED machines [15]. Indeed, in closed loop control, it is possible to exploit different devices to monitor the deposition and melting processes, by using the recorded data as feedback to setup the optimal process parameters.

In this paper, the first part of a larger research project is presented, in view to design a tailored deposition closed loop control to improve the performances of DED systems, guaranteeing their maximum efficiency during the process. In particular, this preliminary study is focused on the performances of a double chamber nozzle solution, controlled through the evaluation of the shape of the powder flow at the exit of the nozzle outlet and the spread of the powder particles on the deposition plane. During experimental tests, the powder flow is monitored with a high-speed camera, in order to confirm its simulating behavior obtained through a Computational Fluid Dynamics CFD model [16,17]. The experimental tests demonstrate that, in the analyzed configurations, the shielding gas seems to have a significant influence on the powder distribution and powder flow geometry.

2. Materials and Methods

A new deposition nozzle solution for DED is designed at the Department of Innovative Technologies (DTI) of University of Applied Sciences and Arts of Southern Switzerland (SUPSI), Manno-CH (see Figure 1a). It consists of a hybrid solution of continuous and discontinuous coaxial systems, with a double chamber coaxial nozzle (see Figure 1b), where the central powder-gas flow is shielded by an external annular flux of inert gas in order to prevent the oxidation of the melt pool, even in the case of no-hermetic deposition chamber, and to improve the catchment efficiency during the deposition, limiting the spread of the powder flow coming out from the nozzle outlet. Two deposition nozzles with an inclination of 30° are placed in a testing chamber employed to characterize the powder flow coming out from the nozzle outlets (see Figure 2a). The deposition nozzles are connected with a flexible feeding system so that the control of shielding, carrier, and powder mass flow can be independent and precise (see Figure 2b).

Figure 1. (a) New nozzle solution assembly; (b) exploded diagram of the nozzle.

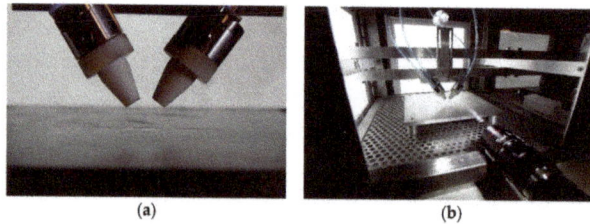

Figure 2. (a) Deposition nozzle configuration; (b) custom feeding system demonstrator at SUPSI.

In order to investigate the performances of the new nozzle solution, an experimental campaign is carried out to analyse the powder spread at 15 mm from the nozzle outlet (i.e., the location of the deposition plane). The chosen carrier and shielding inert gas is argon. The metal powder employed during this experimental investigation is a gas atomized Ti-6Al-4V powder (EOS GmbH-Electro Optical Systems) with a particle size distribution ranging between 45 μm and 105 μm (d_{50}: 70 ± 5 μm, as mentioned by the powder supplier).

As shown in Figure 3, in order to evaluate the performance of the nozzle solution, three different process conditions are considered:

- in the first one, only one nozzle is active without the presence of the substrate (see Figure 3a);
- in the second one, two nozzles are active without the presence of the substrate (see Figure 3b);
- in the third one, two nozzles and a flat substrate placed at 15 mm from the centre of the nozzle outlet are present. In this case, a flat substrate is included in order to analyse the effectiveness of the provided nozzle solution in limiting the spread of the powder particles at the first layer deposition (see Figure 3c).

In particular, experimental tests are carried out with various carrier, shielding, and powder mass flow rates applying a full factorial 2 × 3 × 2 design, according to the Design of Experiments (DOE) approach, as listed in Table 1. Each value range is fixed starting from process parameters commonly adopted in DED [9]. The experimental results are investigated, taking into account a confidence interval of 95% and running three repetitions for each parameter setting.

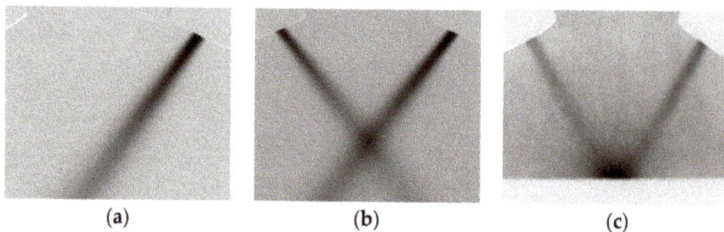

Figure 3. (a) One active nozzle; (b) two active nozzles; (c) two active nozzles with a substrate at 15 mm.

Table 1. Evaluated experimental factors.

Process Parameters for One Nozzle	Low level	Medium level	High level
carrier mass flow rate (kg/s)	$4.51e^{-0.5}$		$5.41e^{-0.5}$
shielding flow rate (kg/s)	0	$1.95e^{-0.4}$	$2.92e^{-0.4}$
powder feed rate (g/s)	0.1		0.14

A high-speed acquisition camera records the variation of the powder flow for every combination of process parameters, and the shape and width variation of the powder flow is detected and measured through image analyses performed with the open software ImageJ [18] (see Figure 4).

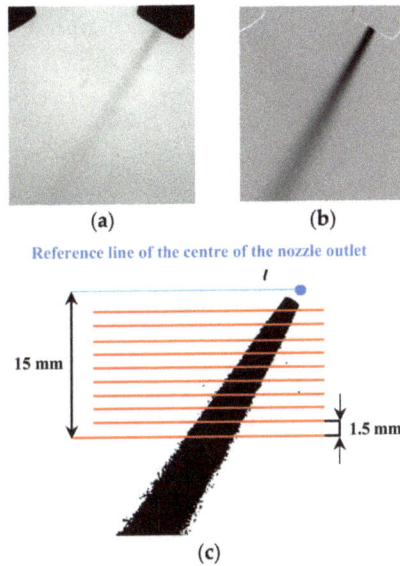

Figure 4. (**a**) Acquired image; (**b**) acquired image after subtracting the background; (**c**) evaluation planes on the image filtered by "binary-mask".

To correctly analyse the powder flow profile, reducing the presence of undesired floating particles which can deteriorate and worsen the powder flow characterization, 10 images are extracted from each experimental video and subsequently averaged (see Figure 4a). From the resulting image, the corresponding background is subtracted (see Figure 4b) and a "binary-mask" filter is applied to highlight the contour and the shape of the powder flow (see Figure 4c). To characterize the width variation of the powder flow along its extension, 10 equally spaced evaluation planes are traced below the nozzle outlet (i.e., the red lines in Figure 4c). The 95% of the portion of plane crossing the powder flow is taken into account as the local width of the powder flow, cutting the total powder flow intensity in correspondence to the analysed evaluation plane of 2.5% at its beginning and end. This is required due to the irregular shape of the powder flow along its edges.

The binary-mask filter converts an image in black and white thanks to the preliminary definition of an intensity threshold. This tool is very useful in image processing, nevertheless, in this case, the choice of an intensity threshold is critical since it directly affects the size and extension of the powder flow (see Figure 5a,b). To objectively determine this value, an iterative method is designed, forcing an equal total intensity of the powder flow through the last five evaluation planes (see Figure 6). The purpose of this constraint is to suppose an equal quantity of powder particles passing through the evaluation plane for an equal value of total intensity in correspondence with each plane. The equation employed in this case is listed below:

$$\frac{\sum_{k=1}^{5} I_k}{5} - \left(\frac{\sum_{k=1}^{5} I_k}{5}\right) 5\% < I_k < \frac{\sum_{k=1}^{5} I_k}{5} + \left(\frac{\sum_{k=1}^{5} I_k}{5}\right) 5\% \; \forall i = 1, \ldots, 5 \tag{1}$$

where I_k is the corresponding total intensity of the powder flow for each evaluated plane. For each experimental video, the intensity thresholds chosen for the binary-mask filter are those satisfying Equation (1) (see Figure 6).

(a)   (b)

Figure 5. Binary-mask filter: resulting powder flow for different intensity thresholds (carrier = $4.51e^{-0.5}$ kg/s; shielding = $2.92e^{-0.4}$ kg/s; powder feed rate = 0.14 g/s): (**a**) intensity threshold of 206; (**b**) intensity threshold of 198.

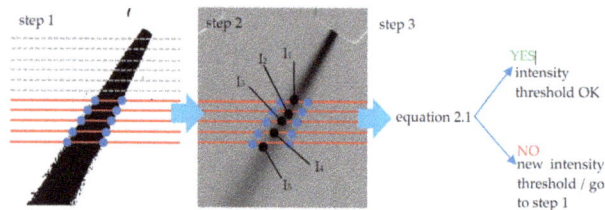

Figure 6. Scheme of the iterative method employed to determine the intensity threshold.

3. CFD Modeling and Theoretical Assumptions

The deposition simulation is approached by using CFD analysis, as generally occurs for DED processes, modelling particle powders as a discrete phase dispersed into a continuous phase (i.e., the inert gas). This approach is widely employed in the scientific literature to simulate fluid-dynamics problems when a low volume fraction of powder particles is dispersed and dragged by a gas. Carrier and shielding gases are computed as a continuous phase applying the standard κ-ε turbulent flow model available in ANSYS® FLUENT (v15.0) and based on the Navier-Stokes equations. To describe a turbulent flow, the time-averaging governing equations are:

- conservation of mass:

$$\frac{\partial}{\partial x}(\rho u_i) = 0 \tag{2}$$

where ρ is the argon density (1.623 kg/m^3), u_i is the gas velocity, and x_i is the gas position.

- Conservation of momentum:

$$\frac{\partial}{\partial x}(\rho u_i u_j) = -\frac{\partial p}{\partial x_i} + \frac{\partial\left(\left[(\mu + \mu_t)\left(\frac{\partial u_i}{\partial x_j} + \frac{\partial u_j}{\partial x_i}\right)\right]\right)}{\partial x_j} + \rho g_i \tag{3}$$

where p is the pressure, g is the gravitational acceleration, μ is the molecular viscosity (2.125×10^{-5} kg/m s), and μ_t is the turbulent viscosity.

The most commonly used model to handle this situation is the standard k-ε model, in which k and ε represent the turbulent kinetic energy and the dissipation of kinetic energy respectively. The conservation of the kinetic energy of turbulence is given by:

$$\frac{\partial(\rho k u_i)}{\partial x_i} = \frac{\partial\left[\left(\mu + \frac{\mu_t}{\sigma_k}\right)\frac{\partial k}{\partial x_j}\right]}{\partial x_j} + G_k + G_b - \rho\varepsilon \tag{4}$$

whereas the conservation of the dissipation of kinetic energy of turbulence is defined as:

$$\frac{\partial}{\partial x_i}(\rho\varepsilon u_i) = \frac{\partial}{\partial x_j}\left[\left(\mu + \frac{\mu_t}{\sigma_\varepsilon}\right)\frac{\partial\varepsilon}{\partial x_j}\right] + C_{1\varepsilon}\frac{\varepsilon}{k}(G_k + G_b) - C_{2\varepsilon}\rho\frac{\varepsilon^2}{k} \tag{5}$$

$$G_k = \mu_t\left(\frac{\partial u_j}{\partial x_i} + \frac{\partial u_i}{\partial x_j}\right)\frac{\partial u_i}{\partial x_j} \tag{6}$$

$$G_b = -g_i\frac{\mu_t}{\rho Pr_t}\frac{\partial\rho}{\partial x_i} \tag{7}$$

where $C_{1\varepsilon} = 1.44$, $C_{2\varepsilon} = 1.92$, $k = 1.0$, and $\varepsilon = 1.3$ are empirical constants; Pr_t is the turbulent Prandtl number; G_k is the generation of turbulence kinetic energy due to the mean velocity gradients; and G_b is the generator of turbulence kinetic energy due to buoyancy.

The equations provided above govern the continuous phase constituted by the argon carrier gas. To complete the theoretical basis of this CFD analysis, the equations governing the secondary phase have to be provided. The discrete phase representing the powder particles dispersed into the continuous phase is computed by ANSYS® FLUENT, which integrates the differential equation of a particle's force balance in a Lagrange coordinate system. The balance of the forces is given by:

$$\frac{du_{p,i}}{dt} = F_D(u - u_p) - g_i\left(\frac{\rho_p - \rho}{\rho_p}\right) + F_i \tag{8}$$

where u_p is the particle velocity, u is the fluid phase velocity, ρ is the fluid density, ρ_p is the density of the particles, g is the gravitational acceleration, and F_i is an additional acceleration (force/unit particle mass) term. The F_D coefficient is the drag force per powder mass unit and it can be calculated as:

$$F_D = \frac{18\mu}{\rho_p d_p^2}\frac{C_D Re}{24} \tag{9}$$

In the equation defining F_D, μ is the molecular viscosity of the fluid, d_p is the particle diameter, Re is the relative Reynolds number, and C_D is the drag coefficient, defined as:

$$Re = \frac{\rho d_p|u_p - u|}{\mu} \tag{10}$$

$$C_D = a_1 + \frac{a_2}{Re} + \frac{a_3}{Re^2} \tag{11}$$

where a_1, a_2, and a_3 are empirical constants. The second term on the right of Equation (8) consists of the gravity and buoyancy forces per unit particle mass. Therefore, the particle velocity can be acquired as:

$$\frac{dx_i}{dt} = u_{p,i} \tag{12}$$

The particle trajectory can be obtained using Equations (8) and (12).

The domain taken into account during the numerical simulation is composed by a control volume large enough to allow the complete development of the powder flow at the exit of the deposition nozzle.

In the case of two active nozzles, a symmetry condition is assumed. The theoretical assumptions taken into account for the numerical analysis are:

- only the forces of drag, inertia, and gravity are included in the analysis;
- collisions among particles are not considered;
- the grain size distribution is considered uniform with an average diameter of 70 μm, since it corresponds to the mass median diameter of the powder particle distribution;
- the gas-powder flow is assumed to be a steady state flow;
- the powder particles are assumed to be spherical in shape;
- the substrate "traps" the powder particles reaching the surface.

4. Results

Table 2 summarizes the results concerning the 95% of the portion of the total powder flow width at 15 mm below the nozzle outlet for the two process conditions where the presence of the substrate is not taken into account (see Figure 3a,b). In the case of only one active nozzle (Figure 3a), the effects of the shielding gas (*p*-value = 0.000) and of the powder feed rate (*p*-value = 0.001) are significant. Increasing the powder feed rate from 0.1 to 0.14 g/s, the 95% of the total spread of the powder flow at 15 mm decreases from 4.58 to 4.43 mm in average, obtaining an average absolute variation of −9.8%. This behaviour is significant and demonstrates the effectiveness of the suggested nozzle solution in the powder particle deposition, since it is capable of ensuring a high particle powder concentration focused at the centre of the flow, thus decreasing the width of the powder flow (i.e., the powder flow width at 15 mm from the nozzle outlet).

The influence of the shielding gas on the powder flow profile is consistent for every evaluated combination of process parameters and it is well represented in Figure 7. The external flow of inert gas compacts the powder particles at the exit of the nozzle and contrasts the effect of the gravity force that tends to pull down the powder particles, deflecting them from the nominal trajectory imposed by the nozzle geometry. Upon increasing the shielding mass flow rate from 0 to $2.92e^{-0.4}$ kg/s, the spread of the powder flow at 15 mm from the nozzle outlet decreases to about −14.6% on average.

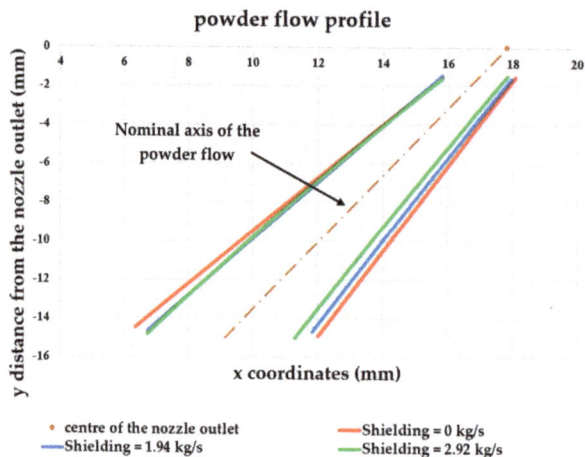

Figure 7. Influence of the shielding gas on the profile of the powder flow (one active nozzle).

Table 2. Summary of the experimental results without the presence of the substrate.

One Active Nozzle	Powder Feed Rate = 0.1 g/s				Powder Feed Rate = 0.14 g/s			
	Carrier = $4.51e^{-0.5}$ kg/s		Carrier = $5.41e^{-0.5}$ kg/s		Carrier = $4.51e^{-0.5}$ kg/s		Carrier = $5.41e^{-0.5}$ kg/s	
	95 % Width (mm)	Variation (%)	95 % Width (mm)	Variation (%)	95 % Width (mm)	Variation (%)	95 % Width (mm)	Variation (%)
Shielding = 0 kg/s	4.93	-	4.98	-	4.77	-	4.97	-
Shielding = $1.95e^{-0.4}$ kg/s	4.33	−12	4.57	−8.3	4.41	−7.6	4.33	−12.8
Shielding = $2.92e^{-0.4}$ kg/s	4.29	−13	4.35	−12.7	4.22	−11.5	3.91	−21.3

Two Active Nozzles	Powder Feed Rate = 0.2 g/s				Powder Feed Rate = 0.28 g/s			
	Carrier = $9.02e^{-0.5}$ kg/s		Carrier = $1.08e^{-0.4}$ kg/s		Carrier = $9.02e^{-0.5}$ kg/s		Carrier = $1.08e^{-0.4}$ kg/s	
	95 % Width (mm)	Variation (%)	95 % Width (mm)	Variation (%)	95 % Width (mm)	Variation (%)	95 % Width (mm)	Variation (%)
Shielding = 0 kg/s	5.73	-	6.34	-	5.27	-	5.56	-
Shielding = $3.9e^{-0.4}$ kg/s	5.11	−10.9	5.22	−17.8	4.40	−16.5	5.68	−2.3
Shielding = $5.84e^{-0.4}$ kg/s	5.29	−7.8	5.58	−12.1	5.32	−1.1	5.64	−1.5

In the case of two active nozzles without the presence of the substrate, analysis of variance (ANOVA) demonstrates that the powder feed rate has a negligible effect (p-value = 0.064) on the spread variation of the powder flow compared to the shielding (p-value = 0.001) and carrier gas (p-value = 0.000). Upon increasing the carrier gas from $9.02e^{-0.5}$ to $1.08e^{-0.4}$ kg/s, the spread of the powder flow increases (see Table 2). This is mainly due to the increase in the carrier inertia that limits the effect of the shielding gas. Moreover, the influence of the shielding gas is not linear, as happens in the process condition with only one active nozzle. In this case, the experimental analysis shows a significant reduction in the powder particles diffusions for medium values of the shielding gas, followed by a critical enlargement at the highest values. For medium values of the shielding gas, in fact, the average 95% of the powder flow width at 15 mm decreases down to −11.8%, whereas for higher mass flow rates the width only decreases down to −5.6%. The reason for this behaviour can be related to the location at which the two powder flows meet each other. Indeed, upon increasing the shielding mass flow rate from $3.9e^{-0.4}$ kg/s to $5.84e^{-0.4}$ kg/s, the zone where the two powder flows meet each other ends up being lower compared to the previous cases, moving away from the nozzle outlet. This phenomenon affects the measurements since for high values of the shielding mass flow, the deposition plane is no longer located at 15 mm from the nozzle outlet.

Concerning the case of two active nozzles with the presence of the substrate at 15 mm from the nozzle outlet, the acquired images are strongly affected by a high concentration of bouncing powder particles that degrade the image analysis, making it impossible to define a reliable image intensity threshold and preventing the application of the image-based method previously discussed. For this reason, only a qualitative analysis is possible for comparing the different spectra of the powder particles distribution in correspondence to the surface of the substrate. Figure 8 illustrates the effect of the shielding gas for constant values of the carrier mass flow rate and the powder feed rate. In the case of no shielding (see Figure 8a), the area with the larger particle mass concentration (i.e., the red one) is wider compared with those represented in Figure 8b,c, in which the shielding gas is $3.9e^{-0.4}$ kg/s and $5.84e^{-0.4}$ kg/s, respectively. In particular, when no shielding is applied, the top edges of the area with the maximum powder concentration are indented, indicating a strong rebound of the powder particles on the metal surface of the substrate. On the contrary, Figure 8b,c show a red area with more defined edges and then a limited rebound of the powder particles. The zone with the higher particle concentration seems to be qualitatively smaller for higher values of the shielding mass flow rate (see Figure 8c). Nevertheless, to have a feedback and a quantitative analysis of the influence of the shielding, carrier, and powder mass flow on the deposition efficiency of the process under this process conditions, further experimental tests are required.

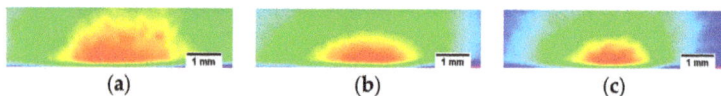

(a)   (b)   (c)

Figure 8. Two active nozzles with a substrate at 15 mm. Carrier mass flow = $1.08e^{-0.4}$ kg/s, powder feed rate = 0.2 g/s: (a) shielding = 0 kg/s; (b) shielding = $3.9e^{-0.4}$ kg/s; (c) shielding = $5.84e^{-0.4}$ kg/s.

5. Discussion

The CFD model is verified, estimating the spread of the powder flow at 15 mm from the nozzle outlet and taking into account the particle mass concentration (see Figure 9a). To correctly compare the numerical outputs with the experimental results, the nominal particle mass concentration is adequately filtered for each combination of process parameters, since:

- the experimental analysis is based on 2D images (as those reported in Figure 4);
- the employment of a "binary-mask" filter requires the application of an intensity threshold to highlight the shape of the powder flow;
- the particle mass concentration is provided by the numerical software with no filtering.

The filtering threshold set to cut off the numerical outputs is equivalent in terms of percentages to the intensity threshold employed during the image analysis. After filtering, 10 lines are transversely traced to discretize the particle mass concentration and detect the spread of the powder flow at 15 mm from the nozzle outlet (as shown in Figure 9b). The numerical powder flow width to compare with the experimental one is estimated by computing the 95% of the total particle mass concentration resulting from the sum of the 10 tracked lines, cutting off the values lower than the 5% of the maximum powder concentration.

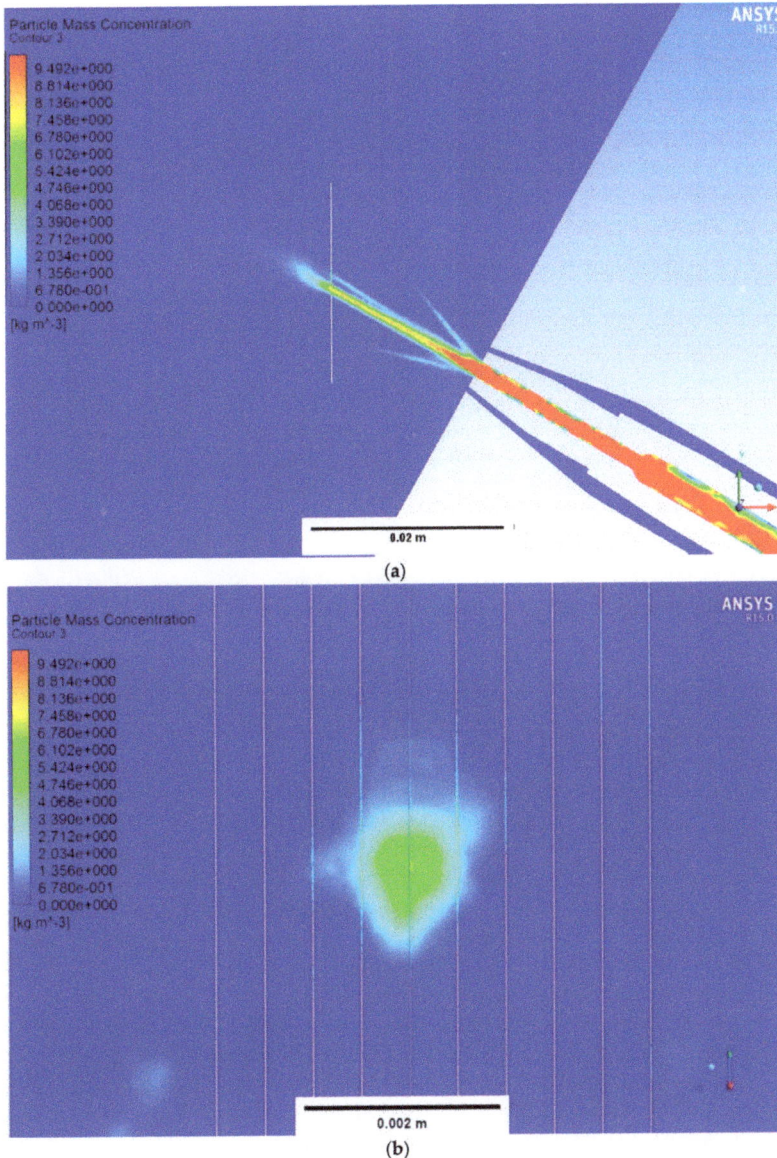

Figure 9. (**a**) Particle mass concentration along the nozzle axis; (**b**) particle mass concentration at the deposition plane and the 10 evaluation lines.

The numerical results do not correspond to the experimental ones; from data reported in Table 3 it is possible to note that the estimated powder flow width is quite different from the results obtained by the experimental characterization.

Table 3. Comparison between experimental and numerical results.

One Active Nozzle	Powder Feed Rate = 0.1 g/s		
	Carrier = $4.51e^{-0.5}$ kg/s		
	Shielding = 0 kg/s	Shielding = $1.95e^{-0.4}$ kg/s	Shielding = $2.92e^{-0.4}$ kg/s
Experimental 95% of the total width [mm]	4.93	4.33	4.29
Numerical 95% of the total width [mm]	5.9	5.5	7.5
Two Active Nozzles	Powder Feed Rate = 0.2 g/s		
	Carrier = $9.02e^{-0.5}$ kg/s		
	Shielding = 0 kg/s	Shielding = $3.90e^{-0.4}$ kg/s	Shielding = $5.84e^{-0.4}$ kg/s
Experimental 95% of the total width [mm]	5.73	5.11	5.29
Numerical 95% of the total width [mm]	4.6	9.4	4.2

For both one active nozzle and two active nozzles systems, the employed CFD model is not capable of correctly estimating the powder behaviour. From an analysis of the CFD results, for each combination of process parameters, the shielding gas seems to have a very low influence on the powder flow geometry (see Figure 10). Indeed, both in the case with no shielding (see Figure 10a) and in the case with shielding (see Figure 10b,c), the graphical CFD results show the presence of secondary isolated powder flows that diverge from the central main powder flow. These secondary flows seem to not be affected by the presence of the shielding that should deflect them or limit their extension.

The motivations for such different results could be attributed to two different reasons:

- the critical issue in the choice of the filtering value to apply to the CFD outputs;
- the inadequacy of the employed CFD model and assumptions.

The employment of a correct value to filter all the data provided by the CFD analysis is a critical issue not easy to deduce, but it is required to compare the numerical results with the experimental ones due to the application of a binary-mask filter during the experimental image analysis. Nevertheless, the numerical method employed in this analysis seems to not fit the behaviour of the powder flow recorded during the experimental investigation, probably because both the influence of the air at the exit of the deposition nozzle is not taken into account and the powder grain size distribution is not correctly estimable by its mass median diameter (i.e., 70 μm). To improve the CFD simulation and to obtain a powder flow behaviour more in compliance with the real one, a more complex model has to be taken into account, such as the Eulerian multiphase model. The Eulerian multiphase is a model implemented in ANSYS® FLUENT, which allows for the modeling of multiple separate, yet interacting phases, where Eulerian treatment is used for each phase, in contrast to the Eulerian-Lagrangian treatment that is used only for the discrete phase model. The application of this numerical model together with a more accurate estimation of the distribution of the powder grain size (i.e., Rosin-Rammler particle size distribution) could improve the experimental fitting of the powder behaviour at the expense of more computational time and memory.

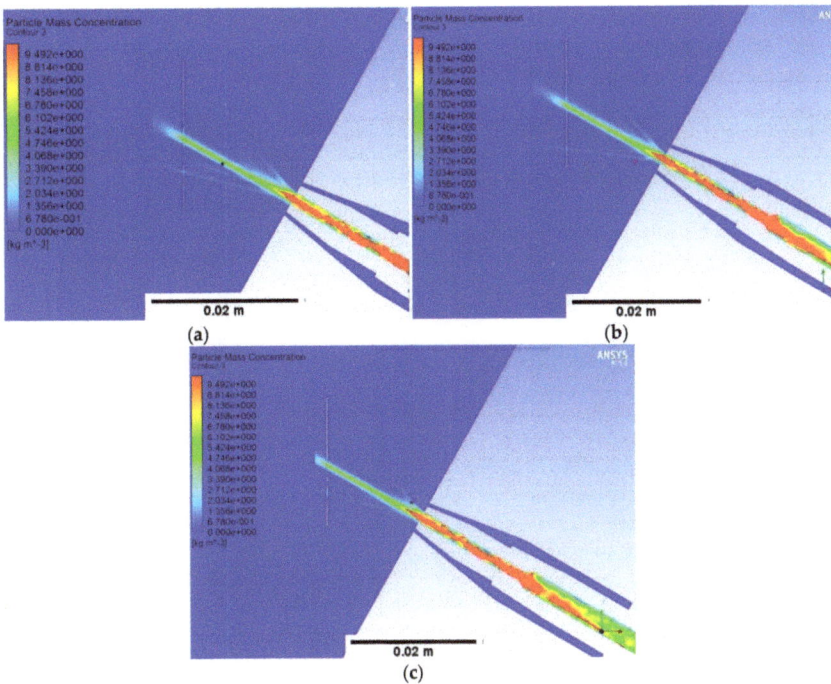

Figure 10. Carrier mass flow rate = $4.51e^{-0.5}$ kg/s and powder feed rate = 0.1 g/s: (**a**) shielding mass flow = 0 kg/s; (**b**) shielding mass flow = $1.95e^{-0.4}$ kg/s; (**c**) shielding mass flow = $2.92e^{-0.4}$ kg/s.

6. Conclusions

In order to optimize the process efficiency of a DED system, the performances of a new deposition nozzle solution designed at DTI of SUPSI (Manno-CH) are monitored through a high-speed camera. The influence of the shielding mass flow rate, carrier mass flow rate, and powder feed rate on the shape of the powder flow at the exit of the nozzle outlet and the spread of the powder particles on the deposition plane is analysed. The main findings of the experimental investigation are:

- the solution of a shielding gas external to the carrier gas significantly affects the powder distribution and powder flow geometry, decreasing the powder spread in correspondence to the deposition plane;
- the external shielding gas contains the spread of the powder particles in opposition to the gravity force and carrier gas inertia that tend to enlarge the powder flow;
- in the case of one active nozzle, increasing the shielding mass flow rate up to $2.92e^{-0.4}$ kg/s leads an average reduction of −14.6% of the powder flow width in correspondence to the deposition plane;
- in the case of two active nozzles with the presence of the substrate, the shielding gas qualitatively seems to reduce the powder rebound, reducing the extension of the zone of higher particle concentration;
- the employed CFD model does not fit the experimental results. A new and more complex theoretical model has to be implemented to simulate the process (i.e., Eulerian multiphase model) providing a more reliable distribution of the powder grain size (i.e., Rosin-Rammler particle size distribution);

Technologies **2017**, *5*, 29

- the results of the experimental campaign highlight that the analysed deposition nozzle can be a good solution for the improvement of the catchment efficiency of a DED system, reducing the powder spread in correspondence to the deposition plane.

Acknowledgments: The authors would like to acknowledge the European research project belonging to the Horizon 2020 research and innovation programme Borealis—the 3A energy class Flexible Machine for the new Additive and Subtractive Manufacturing on next generation of complex 3D metal parts.

Author Contributions: Federico Mazzucato, Mariangela Lombardi and Anna Valente conceived and designed the experiments; Federico Mazzucato performed the experiments; Federico Mazzucato, Simona Tusacciu and Anna Valente analyzed the data; Sara Biamino and Manuel Lai contributed to material selection; Federico Mazzucato, Sara Biamino and Mariangela Lombardi wrote the paper.

Conflicts of Interest: The authors declare no conflict of interest.

1. Dunsky, C. Process monitoring in laser additive manufacturing. *Ind. Laser Solut. Manuf.* **2014**, *29*, 14–20.
2. Sames, W.J.; List, F.A.; Pannala, S.; Dehoff, R.R.; Babu, S.S. The metallurgy and processing science of metal additive manufacturing. *Int. Mater. Rev.* **2016**, *61*, 315–360. [CrossRef]
3. Herzog, D.; Seyda, V.; Wycisk, E.; Emmelmann, C. Additive manufacturing of metals. *Acta Mater.* **2016**, *117*, 371–392. [CrossRef]
4. Scott, J.; Gupta, N.; Weber, C.; Newsome, S.; Wohlers, T.; Caffrey, T. *Additive Manufacturing: Status and Opportunities*; Science and Technology Policy Institute: Washington, DC, USA, 2012; pp. 1–29.
5. Toyserkani, E.; Khajepour, A.; Corbin, S.F. *Laser Cladding*; CRC Press: Boca Raton, FL, USA, 2004.
6. Zekovic, S.; Dwivedi, R.; Kovacevic, R. Numerical simulation and experimental investigation of gas-powder flow from radially symmetrical nozzles in laser-based direct metal deposition. *Int. J. Mach. Tools Manuf.* **2007**, *47*, 112–123. [CrossRef]
7. Ding, Y.; Dwivedi, R.; Kovacevic, R. Process planning for 8-axis robotized laser-based direct metal deposition system: A case on building revolved part. *Robot. Comput. Integr. Manuf.* **2017**, *44*, 67–76. [CrossRef]
8. Fessler, J.R.; Merz, R.; Nickel, A.H.; Prinz, F.B. Laser deposition of metals for shape deposition manufacturing. In *Solid Freeform Fabrication Symposium*; Bourell, D., Beaman, J., Marcus, H., Crawford, R., Barlow, J., Eds.; University of Texas at Austin: Austin, TX, USA, 2017.
9. Thompson, S.T.; Bian, L.; Shamsaei, N.; Yadollahi, A. An overview of Direct Laser Deposition for additive manufacturing; Part I: Transport phenomena, modelling and diagnostics. *Add. Manuf.* **2015**, *8*, 36–62. [CrossRef]
10. Balu, P.; Leggett, P.; Kovacevic, R. Parametric study on a coaxial multi-material powder flow in laser-based powder deposition process. *J. Mater. Process. Technol.* **2012**, *212*, 1598–1610. [CrossRef]
11. Zhu, G.; Li, D.; Zhang, A.; Tang, Y. Numerical simulation of metallic powder flow in a coaxial nozzle in laser direct metal deposition. *Opt. Laser Technol.* **2011**, *43*, 106–113. [CrossRef]
12. Smurov, J.; Doubenskaia, M.; Zaitsev, A. Comprehensive analysis of laser cladding by means of optical diagnostics and numerical simulation. *Surf. Coat. Technol.* **2013**, *220*, 112–121. [CrossRef]
13. Kovaleva, I.; Kovalev, O.; Zaitsev, A.; Smurov, I. Numerical simulation and comparison of powder jet profiles for different types of coaxial nozzles in direct material deposition. *Phys. Procedia* **2013**, *41*, 810–872. [CrossRef]
14. Costa, L.; Vilar, R. Laser powder deposition. *Rapid Prototyp. J.* **2009**, *15*, 264–279. [CrossRef]
15. Song, L.; Bagavath-Singh, V.; Dutta, B.; Mazumder, J. Control of melt pool temperature and deposition height during direct metal deposition process. *Int. J. Adv. Manuf. Technol.* **2012**, *58*, 247–256. [CrossRef]
16. Tabernero, I.; Lamikiz, A.; Ukar, E.; Lopez de Lacalle, L.N.; Angulo, C.; Urbikain, G. Numerical simulation and experimental validation of powder flux distribution in coaxial laser cladding. *J. Mater. Process. Technol.* **2010**, *210*, 2125–2134. [CrossRef]
17. Wen, S.Y.; Shin, Y.C.; Murthy, J.Y.; Sojka, P.E. Modeling of coaxial powder flow for the laser direct deposition process. *Int. J. Heat Mass Transf.* **2009**, *52*, 5867–5877. [CrossRef]
18. ImageJ v1.51k. Available online: https://imagej.nih.gov/ij/ (accessed on 26 May 2017).

technologies

MDPI

Article

Customised Alloy Blends for In-Situ Al339 Alloy Formation Using Anchorless Selective Laser Melting

Pratik Vora [1], Rafael Martinez [1], Neil Hopkinson [1], Iain Todd [2] and Kamran Mumtaz [1,*

[1] Department of Mechanical Engineering, University of Sheffield, Sheffield S1 3JD, UK;
PratikVora@materialssolutions.co.uk (P.V.); ramartinez1@sheffield.ac.uk (R.M.);
nhopkinso@sheffield.ac.uk (N.H.)

[2] Department of Materials Science and Engineering, University of Sheffield, Sheffield S1 3JD, UK;
iain.todd@sheffield.ac.uk

* Correspondence: k.mumtaz@sheffield.ac.uk; Tel.: +44-0114-222-7789

Academic Editors: Salvatore Brischetto, Paolo Maggiore and Carlo Giovanni Ferro
Received: 31 March 2017; Accepted: 19 May 2017; Published: 24 May 2017

Abstract: The additive manufacturing process Selective Laser Melting (SLM) can generate large thermal gradients during the processing of metallic powder; this can in turn lead to increased residual stress formation within a component. Metal anchors or support structures are required to be built during the process and forcibly hold SLM components to a substrate plate and minimise geometric distortion/warpage due to the process induced thermal residual stress. The requirement for support structures can limit the geometric freedom of the SLM process and increase post-processing operations. A novel method known as Anchorless Selective Laser Melting (ASLM) maintains processed material within a stress relieved state throughout the duration of a build. As a result, metal components formed using ASLM do not develop signification residual stresses within the process, thus, the conventional support structures or anchors used are not required to prevent geometric distortion. ASLM locally melts two or more compositionally distinct powdered materials that alloy under the action of the laser, forming into various combinations of hypo/hyper eutectic alloys with a new reduced solidification temperature. This new alloy is maintained in a semi-solid or stress reduced state for a prolonged period during the build with the assistance of elevated powder bed pre-heating. In this paper, custom blends of alloys are designed, manufactured and processed using ASLM. The purpose of this work is to create an Al339 alloy from compositionally distinct powder blends. The in-situ alloying of this material and ASLM processing conditions allowed components to be built in a stress-relieved state, enabling the manufacture of overhanging and unsupported features.

Keywords: additive manufacturing; selective laser melting; alloy design; in-situ alloying

---

1. Introduction

Selective Laser Melting (SLM) is an Additive Manufacturing (AM) process in which layers of metallic powder are selectively melted and fused by a high-powered laser to form fully dense 3D components. The method of layered fabrication, combined with the high precision of laser melting, allows for a greatly expanded design freedom with minimal feedstock waste. SLM is increasingly being used in high value markets to produce various aerospace, automotive and medical components; this is mainly a result of the processes' geometric freedom that is afforded to designers when manufacturing fully dense components from a variety of alloys.

During SLM, a rapid heating/melting of material is followed by a rapid solidification that induces thermal variations across a powder bed; this causes areas of the scanned/processed layer to expand/contract at different rates, subsequently generating residual stress which can cause a component to geometrically distort/warp. Laser based processes (i.e., welding, SLM) are known to introduce large amounts of residual stress, due to the large thermal gradients which are inherently present in the process [1]. The AM process of Electron Beam Melting (EBM) uses a much higher powder bed pre-heating temperature than SLM. As a result of EBM's much slower cooling rate, its components develop lower thermally induced residual stresses than SLM [2]. The amount of thermal residual stress generated during the process varies dependent on geometry, material and processing parameters. Within the EBM process, a method involving the creation of sacrificial solid structures in-situ directly below specific EBM geometries allowed large overhanging and unsupported features to be created. These solid structures or "heat supports" reduced detrimental thermal effects by maintaining the unsupported structure at elevated temperatures and reducing thermal gradients. This process, however, is operating at a much higher bed temperature than SLM (due to the requirement for electron beam powder bed pre-heating) and requires the creation of sacrificial structures that are later disposed [3]. Work has shown that processing parameters can be adjusted within an SLM build to maximize the length of an unsupported overhang. In the work undertaken by Mertens et al. [4] laser power and scan spacing was adjusted when fabricating a horizontal unsupported section of an AlSi10Mg component. However, each end of the large overhanging horizontal face was physically attached to a solid vertical SLM geometry and substrate. The mechanical attachment at both ends of the overhang also altered the heat dissipation within the process and resulting thermal gradients generated at these warp prone sections. Without these vertical attachments, curling was experienced due to residual stress at the ends of the large overhanging unsupported geometries. Studies that have focused on reducing residual stress during an SLM build have found that pre-heating the powder bed was the most effective stress reduction method [5].

Often metallic components formed using SLM require support structures or anchors as shown in Figure 1. Anchors are metallurgically fused to the substrate plate and various locations across the laser melted component, forcibly holding geometries in place. Anchors are made from the same material as the SLM component and are also formed through the layer by layer melting of powder within the powder bed. Typically large overhanging/unsupported geometries built parallel to the powder bed require the most support/anchoring [6]. This requirement for anchors/supports restricts the geometric freedom of the process, and increases material/energy utilisation and post processing operations. Because of the limitations anchors exert over the process, efforts to limit the number of supports/anchors and minimise residual stress remains a major research priority today.

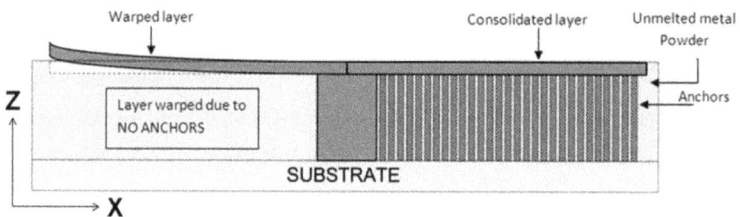

Figure 1. Schematic of un-supported layer susceptible to warp.

1.1. Anchorless Selective Laser Melting

Removing or alleviating stress build up and the requirement for anchors within SLM can be achieved by preventing parts from completely solidifying during processing or maintaining in a stress reduced state. Anchorless Selective Laser Melting (ASLM) or Semi-Solid Processing (SSP) has been developed to prevent processed metal from completely solidifying during an SLM build [7,8].

This is achieved by forming a eutectic alloy or eutectic system (hyper/hypo eutectic) from two or more compositionally distinct materials and maintaining the powder bed pre-heating close to the eutectic melting/solidification point of the newly in process formed alloy. Figure 2 shows a simple binary eutectic phase diagram. Alpha, beta, solid and liquid phases are shown with respect to varying material compositions and temperature; T_E represents the eutectic melting point. Eutectic material proportions can vary from the exact eutectic point creating hypo or hyper eutectic (alloys containing a eutectic system) with variable solidification temperatures and material properties. The range of compositions that have the potential to form eutectics is broad, ranging from aluminium alloys to higher temperature nickels.

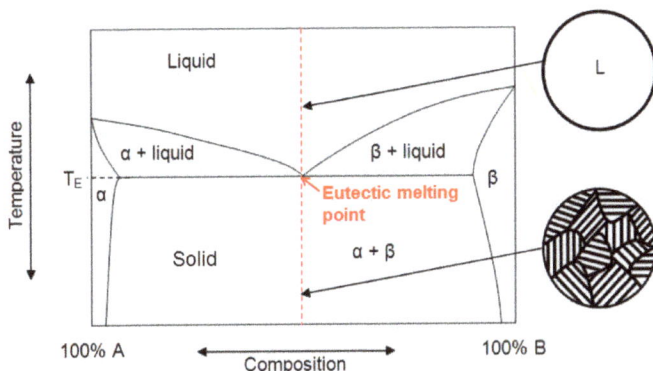

Figure 2. Binary phase diagram containing material A & B.

Figure 3 illustrates the ASLM method and its use of eutectic materials. A batch of material A and B powder is mixed in their un-alloyed eutectic proportions. These materials are then deposited during the ASLM process while maintaining the bed temperature near the eutectic point of the alloy but less than the melt temperature of the individual un-alloyed powder (to prevent melting and agglomeration of un-processed feedstock). It may still be possible to pre-heat the powder bed to temperatures below the eutectic melt point so that stresses are not developed or are sufficiently relieved. Stresses can be sufficiently relaxed if the bed temperature allows diffusional relaxation of the material [9]. Dependent on the material, these relaxation kinetics are initiated at between 40% and 60% of the solidification temperature of the material; this is also time dependent. When the laser scans regions of the powder bed, the individual powders A and B will melt and form a eutectic alloy in-situ; this forms a new solidification temperature that will now only solidify at temperatures below the eutectic solidification point. Since the bed temperature is set near the eutectic point, the melted/alloyed regions will not rapidly solidify or if they are within the diffusional temperature range, they will generate less residual stress than those formed during conventional SLM (rapid melting/solidification rates). Eutectic compositions such as Al66Mg offer large processing windows of 212 °C (temperature difference between eutectic melt point and lowest melting point of individual un-alloyed material). A large processing window may be advantageous as the bed temperature control would not need to be regulated as precisely compared to that of a small processing window. Furthermore, a large processing window may reduce unwanted solid state sintering of unprocessed powders due to pre-heat temperature being far lower than the melt temperature of the un-alloyed material. This solid state sintering or "caking" of material can cause material deposition issues due to powder agglomeration.

Stage 1	Stage 2	Stage 3
Heated powder bed, mix of un-alloyed powders A & B	Laser melts metal A & B forming a new eutectic (or hypo/hyper eutectic) alloy	Bed temperature held at temperature T. Processed material cools uniformly

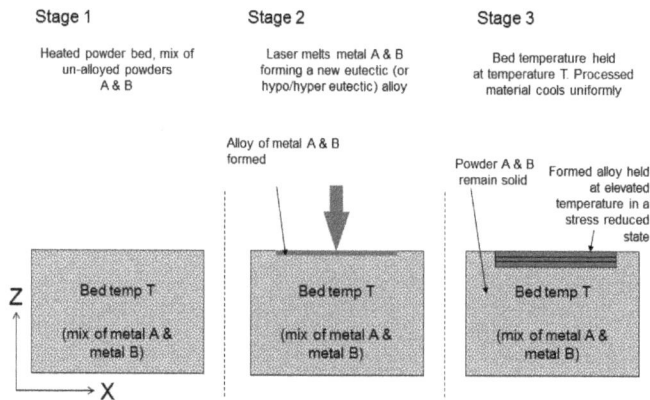

Figure 3. Anchorless selective laser melting methodology.

1.2. Custom Alloy Feedstocks and In-Situ Alloying

The composition of powders used as SLM feedstock are often based upon standard alloys used in conventional manufacturing processes (e.g., casting). These material alloy systems are designed for the application and manufacturing process used. Sourcing custom gas atomised powder feedstock for SLM processing is costly and can inhibit the experimental development of new alloys. Alternative material development methods are typically required to build confidence in a material's process ability/properties before investing in the full scale manufacture of a new alloy powder. This typically would be done by melting batch elements in a furnace under controlled conditions, producing a billet requiring further analysis. Customisation of powders can be used to improve processing ability of materials using SLM or Electron Beam Melting (EBM) technologies. Elemental powder blends and Metal Matrix Composites (MMC) have been processed using SLM technology [10–12]. This has enabled a faster, more cost-effective development of new powder feedstock.

Bartkowiak et al. [13] demonstrated the use of elemental mixtures of Al-Cu powder as feedstock material for the SLM process. The utilisation of elemental mixtures demonstrated a cost-effective approach towards the designing of powder feedstock prior to committing to procure/manufacture new designed powders for research study. In this approach, elemental powder elements were mixed in desired weight percentage and processed under a laser forming an alloy in-situ. In-situ alloying of binary alloys using the copper titanium alloy system (TiCu28) [11]. The purpose of the study was to identify processing parameters for alloying binary alloy systems. The single line tracks melted indicated in-situ alloying of material to a maximum density of 84%. However, it was suggested that the oxidation during melting inhibited densification of melt tracks; the oxygen trapped in the powder feedstock was a primary source of oxidation. This is very common in feedstock powders and therefore, low oxygen content quality powders are often required to overcome this issue. In-situ alloying of elemental/alloy blends is required for the ASLM process to maintain process material in a semi-solid or stress-reduced state, SLM in-situ alloying with blends incorporating Bi-Zn, Al-Si and Ti-Cu have been successfully attempted [7,8,14].

2. Al339 Alloy and Design

The powder feedstock used in conventional SLM would typically be processed in a pre-alloyed state. ASLM requires feedstock to contain at least two components with differing chemical compositions with individually higher melting temperatures compared to their combined/alloys eutectic solidification temperature.

3XX series aluminium casting alloys have been popular for use in industry due to their physical properties and superior cast-ability. They are used widely for applications within automobile, aerospace and medical industries [15]. A screening activity was performed on various aluminium alloys system and aluminium alloy-Al339 was identified as a suitable candidate material due it having a similar Si content to eutectic alloys successfully processed using ASLM.

The design of feedstock material for ASLM processing required the Al339 alloy to be separated into two blends in order to reduce the overlap melting and solidification temperature; i.e., the introduction of super-cooling of material at solidification. This super-cooling behaviour of material is very common in polymer AM [8,16]. This was primarily done by separating elements responsible for eutectic solidification and placing them into separate alloys. A series of combinations (recipes) were formulated and studied for factors such as number of elements in a single alloy, number of pre-alloyed powder batches, mixing ratio, and the processing window generated. The processing window is the difference between the lowest melting temperature of any of the un-alloyed batches within the feedstock used in the blend and the solidification temperature of in-situ alloyed material after processing.

Based on numerous iterations within Thermo-Calc (Version 1.6), the alloy combination with the largest processing window was used. Al339 alloy was split into two pre-alloyed batches, Al-Mg and Si-Cu-Ni. The mixing ratio for Alloy A:Alloy B was 84.75:15.25; this was approximated to 85:15. The composition of the two pre-alloyed batches are shown in Table 1. Individually, alloy A and alloy B do not constitute an Al339 alloy, however, when weighted/mixed in specified proportions (alloy A + alloy B) and fully alloyed together under the action of the laser, they will form the Al339 alloy. The processing window is the difference between the lowest solidus temperature of the powder batch (alloy A or B) and the solidus temperature of the alloy.

Table 1. Alloy feedstock mixing ratio and composition (alloy A + alloy B = Al339).

Elements	Mixing Ratio (A:B)	Solidus Temperature (°C)	Liquidus Temperature (°C)	Processing Window (°C)	Al	Si	Cu	Mg	Ni
Alloy A (wt.%)	85	628	650		98.8	-	-	1.2	-
				51					
Alloy B (wt.%)	15	798	1350		-	78.7	14.8	-	6.5
Pre-alloyed Al339 (wt.%)	-	577	660	-	83.75	12	2.25	1	1

In conjunction with theoretical modelling, Thermo-Calc was used to identify some of the key properties of alloy chemistry that assisted in calculating the processing window and also enabled the identification of theoretical phases in the designed alloy (shown in Figure 4).

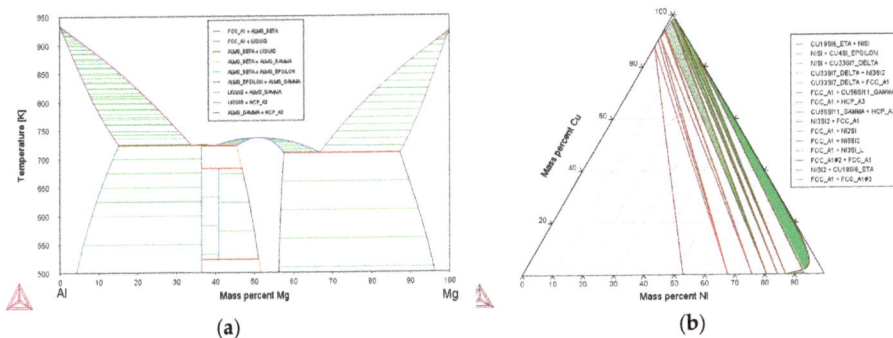

(a) (b)

Figure 4. Phase diagram alloy A (**a**) and alloy B (**b**) using Thermo-Calc.

3. Experimental Procedure

3.1. SLM Processing and Optimisation

A Renishaw SLM 125 system with a 200W fibre laser was used for this study. The scan speed of the laser was derived from point distance (μm, linear distance between two laser exposure spots) and exposure time (μs, duration of exposure of laser spot). The process chamber was fitted with a 125 mm × 125 mm × 100 mm build volume as is standard. However, for this study, a custom high temperature substrate heater was designed and integrated into the SLM 125 system. The designed bed was capable of heating the powder bed up to 380 °C, but in doing so it also reduced the build volume size to Ø 50 mm × 40 mm.

Process parameter optimization for blends of powders was undertaken at three bed temperatures: room temperature, 100 °C and 380 °C. For this, a series of cubes 5 mm × 5 mm × 5 mm were produced and analysed for density. For high throughput, the laser power was set at a maximum of 200 W and parameters such as point distance, exposure time and hatch spacing were altered; a layer thickness of 40 μm was used. Laser Energy Density (ED) is the total energy inputted by the laser onto the powder bed to melt the powder and is a function of laser power (W), relative scan speed (mm/s) and hatch spacing. For heated bed trials, prior to processing at higher temperatures, the powder bed was allowed to soak in heat for 30 min.

3.2. Powder Preparation and Mixing

Gas atomised custom designed alloy powders (A and B) and a Al339 pre-alloyed powder in the range of 15 μm to 45 μm were procured for this study. The powder morphology is shown in Figure 5 for both the blended (alloys A and B) and pre-alloyed Al339 powder. The powder particle size distribution was controlled by sieving the virgin powder using a 63 μm sieve. This resulted in improved control over powder particle size distribution and improved flowability during deposition. To formulate the correct blend of feedstock powder for in-situ alloy of alloy A + B, the powders were weighed and placed in a container with mixing media. A planetary mixer (Speed mixer DAC 800) was used to blend powder batches up to 700 g. The mixer was capable of mixing powder blends at a variable speed of 800–1950 rpm for 5 s to 10 min. The powder blend was mixed at 1000 rpm for three cycles of 2 min each. Zirconia mixing balls of various sizes (large, medium and small) were used as a mixing aid to break the agglomeration of powder and thus enabled effective powder mixing.

Figure 5. SEM image of powder feedstock (**a**) Alloys A (AlMg) + B (SiCuNi) blended (**b**) Al339 pre-alloyed.

3.3. Sample Analysis

The processed samples were cut across the XZ plane (vertically) for microstructural analysis. The samples were processed in accordance to ASTM standard E407-07. Optical and electron microscopy were carried out on the mounted/polished samples to analyse the density and microstructure of consolidated material. Nikon light optical microscopes were used to capture images and the software ImageJ (Version 1.6) was used to measure the density/porosity of parts. An electron microscope was used to observe the microstructure of the samples and phase composition analysis was performed using Siemens D-5000 X-Ray Diffraction (XRD).

As a measure of ASLM stress reduction capability, a warp measurement was carried out on "T" shaped test samples built as part of this study. The sample design was chosen due to the requirement for adding supports to overhanging surfaces during conventional SLM processing; measuring geometric distortion is an efficient methodology for quickly quantifying the level of residual stress within SLM components [14,17]. As the scanning laser processes and melts the powder, the top surface will then cool and shrink; this causes the layer to warp (curl) upwards during the consolidation process. An optical method was employed to measure part distortion. The part would be captured optically (Olympus optical microscope) and images were analysed using Omnimet software to measure warp, as shown in Figure 6. Software was calibrated using scale bars. The warp measurement was expressed in linear distance between horizontal baselines to the most extreme point of a warp surface.

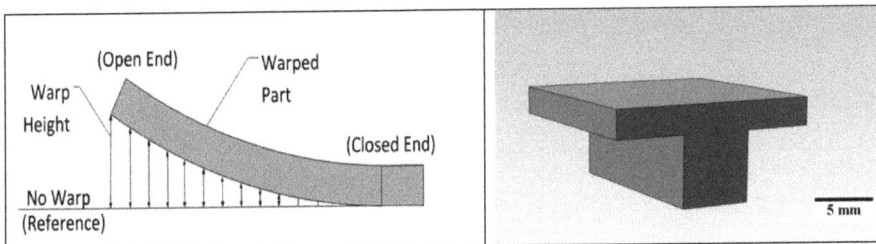

Figure 6. Warp (curling) measurement for "T" shaped components.

4. Results

4.1. Parameter Optimisation

Processing parameter optimisation was carried out at room temperature for blended powders (Alloy A + Alloy B) (100 °C and 380 °C bed temperatures for blended powder using a maximum laser power of 200 W and variable laser exposure time and point distance). Figure 7a shows the relative density of blended powder fabricated samples as a function of Energy Density (ED) in J/mm^2. At room temperature, a maximum density of 99.5% was obtained using a blended powder mixture. As bed temperature increased, the energy required to transform solid powder into liquid and thus consolidate material reduced. As expected, the ED has an inverse relationship to bed temperature. Figure 7b shows a reduction in ED to achieve approximately 99% density at a 380 °C bed temperature. It also reveals that excessive energy inputted to melt the powder bed results in evaporation of material causing additional porosity within the processed material. At a bed temperature of 380 °C, an ED higher than 8.2 J/mm^2 resulted in a relative density of samples that were lower than 95%. A relative density of 98.5% was achieved at 5.64 J/mm^2 at 380 °C. While processing at room temperature or an elevated temperature of 100 °C, the ED requirement was very similar. This suggests that bed temperature had no significant effect at 100 °C. It was also noticed that the ED required to process Al339 at 380 °C was almost half of that required at room temperature. These findings suggest that, as expected, the energy input should be less while processing with the assistance of a heated bed.

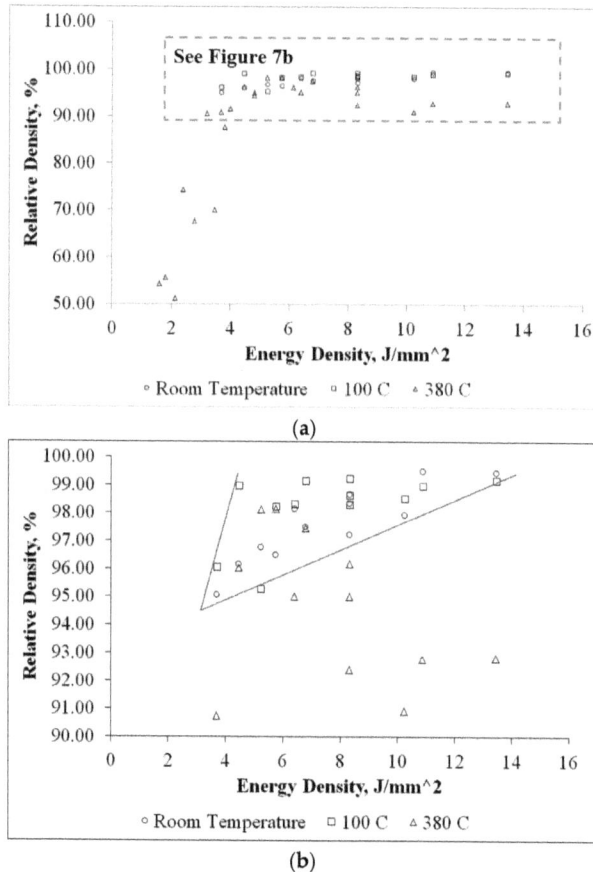

Figure 7. (a) Relative Density (%) as function of Energy Density (J/mm^2) for Blend (Alloy A + Alloy B). (b) Specific region highlighted in Figure 7a.

4.2. Microstructural and Chemical Analysis

Figure 8 shows SEM images of ASLM processed alloy A + B blended feedstock material (forming Al339 in-situ). The fine distribution of the aluminium solid solution and other intermetallic compounds across the melt pool suggested rapid cooling of the material. The primary microstructure observed was α-Al in the form of dendrites and Al-Si eutectics. Upon further magnification, the direction of dendrites suggested the presence of directional solidification that is typical of metal AM microstructures. The melt pools observed in Figure 8 are of different sizes as the laser scan pattern used was a continuous raster scan rotating at 67 deg. The AlCu-θ phase appears to have precipitated around the melt pool boundary, suggesting fine precipitation and solidification at the end. Furthermore, aluminium rich and depleted regions are observed within the same microstructure. This suggested the mixing of powder was relatively uniform with some agglomerations. It is believed that due to the nature of the powder processing (powder blending to layering on powder bed), there would be segregation within the blend that reduces the uniformity of powder feedstock mixing while processing. The microstructure of pre-alloyed Al339 powder processed under the same SLM conditions as the blended powder (100 °C) is shown in Figure 9.

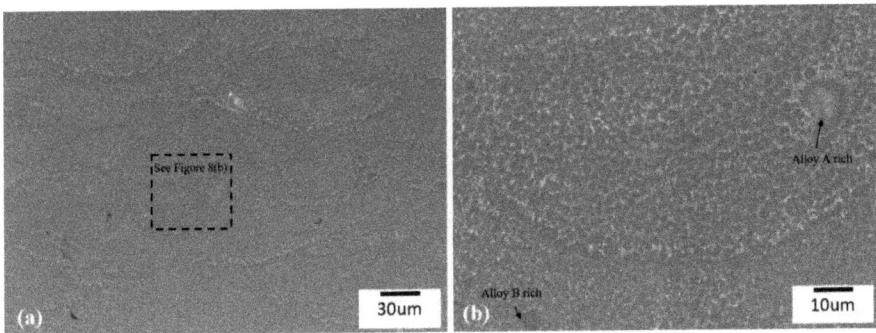

Figure 8. (**a**) SEM Image of Anchorless Selective Laser Melting (ASLM) processed alloy A(AlMg) + B(SiCuNi) blend (85 wt.% alloy A and 15 wt.% alloy B) forming Al339 at 100 °C. (**b**) Specific region highlighted in Figure 8a.

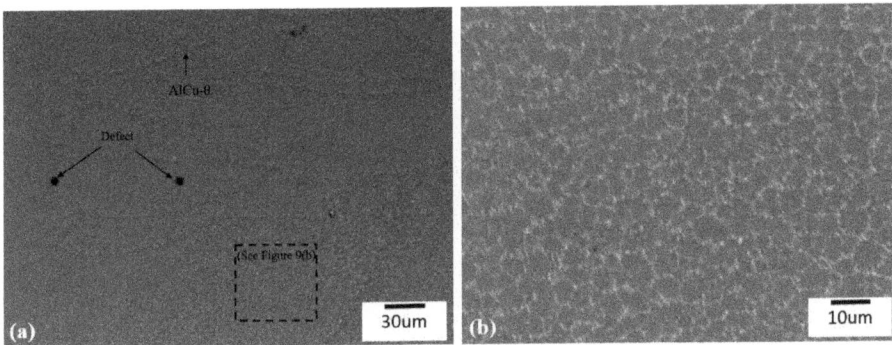

Figure 9. (**a**) SEM image of Al339 pre-alloyed Selective Laser Melting (SLM) sample processed at 100 °C. (**b**) Specific region highlighted in Figure 9a.

Figure 10 shows the microstructure of A + B blended feedstock material (forming Al339 in-situ) at an elevated bed temperature of 380 °C. The processed sample was allowed to cool from 380 °C to room temperature over 3–4 h. The slower cooling resulted in reticular network of α-Al solid solution. In Figure 10, the eutectics structures are denoted by the dark grey areas, the primary constituent of Al339. However, the intermetallic structures are potentially concentrated in the interdendritic spaces; alloy B rich regions are in light grey. In the blended SLM samples, intermetallic phases of AlCu-θ, Al_3Ni, and Mg_2Si are present in minor quantities due to the low composition of the elements, as shown in Figure 11. Directly processing Al339 pre-alloyed powder at 380 °C was not possible due to powder deposition issues (agglomeration of powders due to high powder bed pre-heating; this is discussed further in Section 4.3).

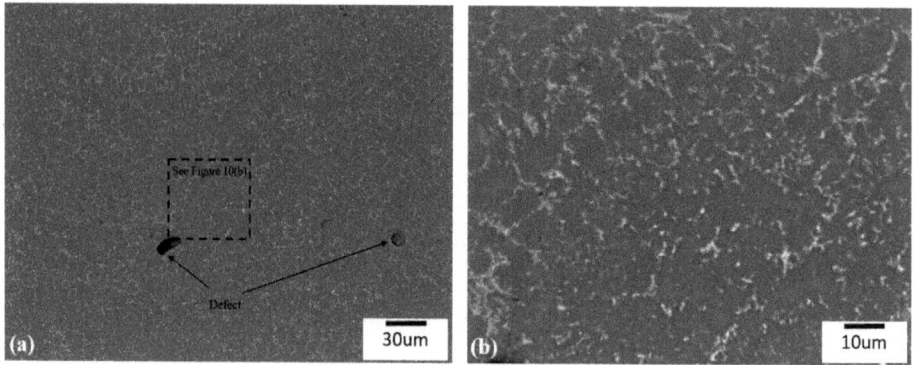

Figure 10. (a) SEM image of Al339 blend (85 wt.% alloy A and 15 wt.% alloy B) processed at 380 °C. (b) Specific region highlighted in Figure 10a.

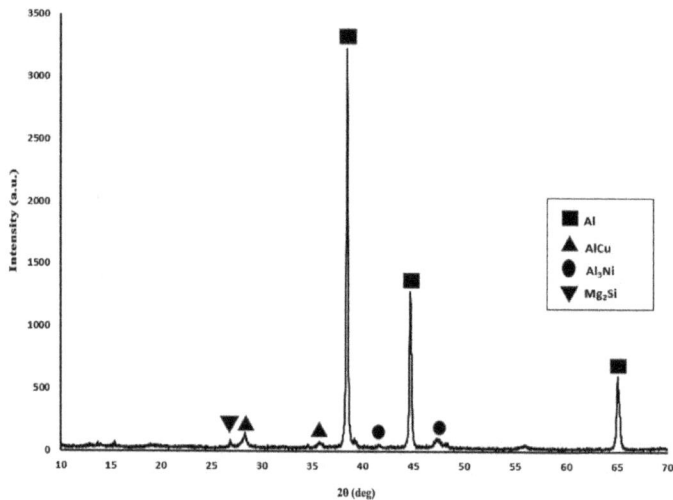

Figure 11. XRD patterns of the Al339 blend processed by SLM.

4.3. Residual Stress Measurement

In this study, the residual stress measurement was undertaken by measuring geometric distortion of the fabricated un-supported overhang geometries (shown in Figure 6). Figure 12 shows the test sample produced using ASLM with two unsupported overhangs. The underside of the overhanging section had satellites formation (not fully melted powder particles attached) increasing surface roughness. An optimisation of the lower skin parameters at a later stage may enable a reduction in satellite formation. The warp observed with this sample was less than 0.1 mm across its 10 mm length and not significant enough to cause builds to fail. It was only possible to successfully produce this geometry at a 380 °C bed temperature. Attempts were made to build this component at lower temperatures (i.e., room temperature and 100 °C), but there was a failure to produce parts due to excessive warpage causing the powder depositor to collide with the sample during processing. Benchmarking studies have been undertaken using similar geometries to determine the maximum unsupported overhang possible using conventional SLM processing of aluminium alloys with similar Si content [14]. It was found that overhangs of no more than 2 mm could be built before significant warpage would occur. It was also found that attempting to directly process pre-alloyed powders at elevated temperatures (380 °C) to produce the overhang component shown in Figure 12 resulted in partial agglomeration of powder particles during deposition (specifically around the larger cross-sections of the overhang structure; this is possibly due to increased heat build-up). This made it difficult to consistently deposit material due to a partial sintering or caking of unprocessed powders leading to build failures. Such problems did not exist with processing of the 5 x 5 mm cubes with blends or pre-alloyed powders due to the smaller surface area processed and a lower heat build-up within the powder bed. The processing of powder blends suffered less from powder agglomeration due to higher melt temperatures compared to the pre-alloy Al339 (alloy A-628 °C alloy B-798 °C and pre-alloyed Al339-577 °C). Issues associated with agglomeration of pre-heated pre-alloyed powders has been experienced in other ASLM work [8].

Studies have shown an inverse relationship between elevated bed temperature and residual stress. By increasing the bed pre-heat temperature, the temperature gradient reduces between the melt-pool and the solidified material leading to a reduction in residual stress build up [18]. The bed temperature used within this study is lower than the solidification temperature of the newly formed alloy (380 °C and 577 °C, respectively). However, during ASLM, the input from the laser locally heats powder surrounding the melt pool which increases the powder bed temperature. This combined with the 380 °C substrate pre-heating would have allowed the processed materials to be held within a semi-solid state for a prolonged period before solidifying compared with a standard SLM process [19]. Additionally the powder bed is maintained within the materials diffusional temperature range (annealing temperature) which in turn promotes a relaxation of stresses, such that significant geometric distortion would not occur [8].

Figure 12. ASLM overhang component processed from a blend of alloy A(AlMg) + B(SiCuNi) to form in-situ Al339.

5. Conclusions

This work designed and processed two pre-alloyed batch blends, AlMg and SiCuNi, to create an Al339 alloy in-situ using ASLM. ASLM processing was able to sufficiently reduce residual stresses developed within overhang geometries, allowing them to be manufactured without anchors/supports.

The process parameter optimisation study for the designed Al339 blended feedstock material was performed at increasing bed temperatures. The influence of bed temperature was marginal up to 100 °C but was more significant at an elevated temperature of 380 °C. The required laser energy density required to melt the material was reduced by approximately half (5.64 J/mm^2) at 380 °C bed temperatures. Processing alloy A(AlMg) + B(SiCuNi) blended powders at 380 °C enabled the production of un-supported overhanging geometries from a complex alloy system. The observed microstructure of in-situ alloyed samples in comparison to pre-alloyed Al339 displayed similarities, however, the in-situ alloyed samples had aluminium rich and depleted regions within the analysed regions, suggesting that segregation in the powder feedstock may have arisen. It is believed this may have occurred due to the method by which powder is deposited when layering of powder on a build plate. However, this segregation was observed to be minimal. Upon analysing overhang geometries made with in-situ alloyed feedstock, overhangs up to 10 mm could be fabricated with less than 0.1 mm geometric distortion as a result of reduced residual stress when using elevated build temperatures. The capability of restructuring a complex aluminium alloy system such as Al339 (Al-Si-Cu-Mg-Ni) by designing pre-alloyed batches of material that later result in a parent alloy when melted under a laser has been demonstrated and opens up the potential for small scale development of new alloys.

Acknowledgments: The authors would like to thank EPSRC for their support during this research (grant number EP/I028331/1).

Author Contributions: Pratik Vora and Rafael Martinez undertook experiments, analysis and contributed to the writing of this work. Iain Todd, Neil Hopkinson and Kamran Mumtaz supervised this investigation and contributed to the writing of this work.

Conflicts of Interest: The authors declare no conflict of interest.

1. Mercelis, P.; Kruth, J.P. Residual stresses in selective laser sintering and selective laser melting. *Rapid Prototyp. J.* **2006**, *12*, 254–265. [CrossRef]
2. Sochalski-Kolbus, L.M.; Payzant, E.A.; Cornwell, P.A.; Watkins, T.R.; Babu, S.S.; Dehoff, R.R.; Lorenz, M.; Ovchinnikova, O.; Duty, C. Comparison of Residual Stresses in Inconel 718 Simple Parts Made by Electron Beam Melting and Direct Laser Metal Sintering. *Metall. Mater. Trans. A* **2015**, *46*, 1419–1432. [CrossRef]
3. Chou, Y.S.; Cooper, K. Systems and Methods for Designing And Fabricating Contact-Free Support Structures for Overhang Geometries of Parts in Powder-Bed Metal Additive Manufacturing. U.S. Patent 14/276.345, 13 November 2014.
4. Mertens, R.; Clijsters, S.; Kempen, K.; Kruth, J.-P. Optimization of Scan Strategies in Selective Laser Melting of Aluminum Parts With Downfacing Areas. *J. Manuf. Sci. Eng.* **2014**, *136*. [CrossRef]
5. Kruth, J.-P.; Deckers, J.; Yasa, E.; Wauthlé, R. Assessing and comparing influencing factors of residual stresses in selective laser melting using a novel analysis method. *Proc. Inst. Mech. Eng. Part B J. Eng. Manuf.* **2012**, *226*, 980–991. [CrossRef]
6. Vandenbroucke, B.; Kruth, J.P. Selective laser melting of biocompatible metals for rapid manufacturing of medical parts. *Rapid Prototyp. J.* **2007**, *13*, 196–203. [CrossRef]
7. Mumtaz, K.; Vora, P.; Hopkinson, N. A method to eliminate anchors/supports from directly laser melted metal powder bed processes. In Proceedings of the Solid Freeform Fabrication, Austin, TX, USA, 8–10 August 2011.
8. Vora, P.; Mumtaz, K.; Todd, I.; Hopkinson, N. AlSi12 in-situ alloy formation and residual stress reduction using anchorless selective laser melting. *Addit. Manuf.* **2015**, *7*, 12–19. [CrossRef]
9. Onaka, S.; Okada, T.; Kato, M. Relaxation kinetics and relaxed stresses caused by interface diffusion around spheroidal inclusions. *Acta Metall. Mater.* **1991**, *39*, 971–978. [CrossRef]
10. Kühnle, T.; Partes, K. In-Situ Formation of Titanium Boride and Titanium Carbide by Selective Laser Melting. *Phys. Procedia* **2012**, *39*, 432–438. [CrossRef]
11. Sanz-Guerrero, J.; Ramos-Grez, J. Effect of total applied energy density on the densification of copper–titanium slabs produced by a DMLF process. *J. Mater. Proc. Technol.* **2008**, *202*, 339–346. [CrossRef]
12. Attar, H.; Bönisch, M.; Calin, M.; Zhang, L.-C.; Scudino, S.; Eckert, J. Selective laser melting of in situ titanium–titanium boride composites: Processing, microstructure and mechanical properties. *Acta Mater.* **2014**, *76*, 13–22. [CrossRef]
13. Bartkowiak, K.; Ullrich, S.; Frick, T.; Schmidt, M. New Developments of Laser Processing Aluminium Alloys via Additive Manufacturing Technique. *Phys. Procedia* **2011**, *12*, 393–401. [CrossRef]
14. Mumtaz, K.; Hopkinson, N.; Stapleton, D.; Todd, I.; Derguti, F.; Vora, P. Benchmarking metal powder bed Additive Manufacturing processes (SLM and EBM) to build flat overhanging geometries without supports. In Proceedings of the Solid Freeform Fabrication Symposium, Austin, TX, USA, 6–8 August 2012.

15. Baker, H.; Okamoto, H. Alloy Phase Diagrams. In *ASM Handbook*; ASM International: Materials Park, OH, USA, 1992; Volume 3.

16. Kruth, J.P.; Levy, G.; Schindel, R.; Craeghs, T.; Yasa, E. Consolidation of polymer powders by selective laser sintering. In Proceedings of the International Conference on Polymers and Moulds Innovations, Gent, Belgium, 17–19 September 2008; pp. 15–30.

17. Yadroitsava, I.; Grewar, S.; Hattingh, D.; Yadroitsev, I. Residual Stress in Slm Ti6al4v Alloy Specimens. *Mater. Sci. Forum* **2015**, *828*, 305–310. [CrossRef]

18. Ali, H.; Ma, L.; Ghadbeigi, H.; Mumtaz, K. In-situ residual stress reduction, martensitic decomposition and mechanical properties enhancement through high temperature powder bed pre-heating of Selective Laser Melted Ti6Al4V. *Mater. Sci. Eng. A* **2017**, *695*, 211–220. [CrossRef]

19. Brandl, E.; Heckenberger, U.; Holzinger, V.; Buchbinder, D. Additive manufactured AlSi10Mg samples using Selective Laser Melting (SLM): Microstructure, high cycle fatigue, and fracture behavior. *Mater. Des.* **2012**, *34*, 159–169. [CrossRef]

![technologies](technologies logo)

MDPI

Article

Testbed for Multilayer Conformal Additive Manufacturing

Michael D. M. Kutzer * and Levi D. DeVries

Weapons & Systems Engineering, United States Naval Academy, 105 Maryland Avenue,
Annapolis, MD 21402, USA; devries@usna.edu
* Correspondence: kutzer@usna.edu; Tel.: +1-410-293-6113

Academic Editors: Salvatore Brischetto, Paolo Maggiore and Carlo Giovanni Ferro
Received: 8 April 2017; Accepted: 19 May 2017; Published: 24 May 2017

Abstract: Over the last two decades, additive manufacturing (AM) or 3D printing technologies have become pervasive in both the public and private sectors. Despite this growth, there has been little to no deviation from the fundamental approach of building parts using planar layers. This undue reliance on a flat build surface limits part geometry and performance. To address these limitations, a new method of applying material onto or around existing surfaces with multilayer, thick features will be explored. Prior work proposes algorithms for defining conformal layers between existing and desired surfaces, however this work does not address the derivation of deposition paths, trajectories, or required hardware to achieve this new type of deposition. This paper presents (1) the derivation of deposition paths given a prescribed set of layers; (2) the design, characterization, and control of a proof-of-concept testbed; and (3) the derivation and application of time evolving trajectories subject to the material deposition constraints and mechanical constraints of the testbed. Derivations are presented in a general context with examples extending beyond the proposed testbed. Results show the feasibility of conformal material deposition (i.e., onto and around existing surfaces) with multilayer, thick features.

Keywords: additive manufacturing; articulated robotics; coordinated manipulation; coordinated trajectory planning; manipulator control

---

1. Introduction

In a 2015 briefing on AM technologies [1], it was reported that the Naval Systems Engineering Directorate (NAVSEA 05) currently supports upwards of 130 pieces of AM equipment enabling more than a half dozen printing methodologies in materials ranging from ABS plastics to 17-4 PH steel. This equipment is primarily used for research, design, and prototyping applications; however the vision of NAVSEA 05 is to operationalize AM technology in direct support of the fleet with the stated goal of "establish[ing] the processes, specifications and standards for use of AM for ship acquisition, design, maintenance, and operational support [1]". Five years earlier, the Chief Scientist of the Air Vehicle Engineering Department within the Naval Air Systems Command (NAVAIR) [2] identified metallic AM as having the potential "to enhance operational readiness, reduce total-ownership-cost, reduce energy consumption, and enable parts-on-demand manufacturing". In April 2015, NAVAIR reported plans to introduce a flight-critical metal component produced using metallic AM by November of 2017 [3].

Additive manufacturing (AM) or 3D printing technology leverages a variety of processes to bind materials, creating solid structures. AM fabrication offers relaxed design rules and simple part-by-part customization. Unlike part fabrication using subtractive machining, inexperienced developers can produce physical hardware almost immediately with AM; while experienced designers can create complex parts tailored for specific applications. In the context of defense logistics, AM has the capability of "truncating the entire [supply] process and meeting the need exactly where it is [4]". The AM production of certified, field-ready hardware can move the entire supply chain forward to the point of need. This capability can also completely eliminate the need for stockpiled parts, as replacements can be stored electronically and produced on demand. Table 1 summarizes current commercial AM technologies by process (processes are defined using the ASTM F2792 12a Standard Terminology for Additive Manufacturing Technologies).

Table 1. Summary of Commercial AM Technologies [5].

Process	Description	Material(s)
Binder Jetting	A liquid bonding agent is selectively deposited to join powder materials	Polymers, Sand, Glass, Metals
Direct Energy Deposition	Focused thermal energy is used to fuse materials by melting as they are deposited	Metals
Material Extrusion	Material is selectively dispensed through a nozzle or orifice	Polymers
Material Jetting	Droplets of build material are selectively deposited	Polymers, Waxes
Powder Bed Fusion	Thermal energy selectively fuses regions of a powder bed	Metals, Polymers
Sheet Lamination	Sheets of material are bonded to form an object	Paper, Metals
Vat Photopolymerization	Liquid photopolymer in a vat is selectively cured by light-activated polymer	Photopolymers

In AM processes, parts are made by iteratively adding layers of material. Each layer is defined by a thin cross-section of a 3D part exported from a computer-aided design (CAD) model [6]. In general, commercial AM systems use a "build-bed" that serves as the flat substrate for part fabrication. The CAD model is imported into an AM software package, and positioned/oriented relative to the build-bed. Layers are then defined by equally spaced planar slices of the CAD model, parallel to the build-bed. This layering approach is effective for a wide variety of part geometries, however issues may arise with overhanging features. These issues are generally avoided by adding sacrificial support material that is removed following the completion of the AM process [6]. Use of support material (also known as support structure) is common practice in material extrusion, material jetting, and powder bed fusion processes. Although effective, this approach wastes material and adds to the fabrication time. Extensions of the work presented in this paper may reduce the need for support material by coordinating the position and orientation of both the build-bed and print-head.

In this paper, we derive deposition trajectories by coordinating the position and orientation of both the tool-head and build-bed. Similar research has explored the concept of conformal printing onto non-planar surfaces (i.e., surfaces not adhering to the constraints of a plane in Euclidean space) for a variety of applications including subtractive processes like lithography used to produce optics [7,8], and additive methods to fabricate antennas and electronics onto/into mechanical components [9,10]. In general, the AM techniques explored for conformal applications involve "direct write" technologies [6] used to produce thin features on surfaces. One primary exception is the work of Davis et al. [5] that explores algorithms for deriving layers between surfaces, but does not address the transition from layering to material deposition. Related metal deposition methods involve using directed energy and welding to extend printing capabilities by increasing the total degrees of freedom (DOF) or axes used to drive the nozzle or AM tool-head. Unlike the traditional three axis Cartesian (i.e., x, y, and z) stages used by common AM methods like fused deposition modeling (FDM); directed energy deposition (DED) approaches often use four or five axes to drive the relative position between the AM tool-head and build surface (Loughborough University [11]), however deposition generally takes place on a flat or near-flat build-bed.

Conformal AM with multilayer, thick features can be achieved using the layering algorithms presented in [5] in conjunction with registration and manipulation methods commonly used in robotics. This paper replaces the concept of a build-bed with a largely arbitrary "build-object" referring to an application substrate with arbitrary geometry. AM material is directly deposited onto a build-object and layered, adding features to existing surfaces or completely encapsulating the build-object (i.e., applying material to/around an existing part). With sufficient articulation of the build-object, the need for support material commonly used in extrusion-based systems (e.g., FDM) beneath overhanging surfaces [6] may be reduced or eliminated by actively reorienting the deposition path relative to gravity. Doing so may reduce printing time, reduce wasted material, and further reduce design constraints; permitting features such as large sealed cavities. This paper focuses on the derivation of coordinated trajectories for the tool-head and build-object to create prescribed layer geometries. Given the level of articulation redundancy in the proposed system, actively reorienting the deposition direction and build-object orientation relative to gravity to eliminate the need for support material may be attainable but is outside the scope of this work.

While the complexity to implement conformal AM will vary largely with the AM technique, the underlying approach will remain consistent. Assuming the desired (final) part geometry is provided (i.e., produced by a designer in CAD software); this approach requires the following steps:

1. Identify and fixture the build-object,
2. Create (e.g., using 3D scanning) or import a surface model of the build object,
3. Register build-object to a common reference frame,
4. Register desired (designed) part geometry to a common reference frame,
5. Generate the build layers,
6. Generate the tool-head path relative to the build-object adhering to prescribed deposition and system constraints,
7. Generate the tool-head trajectory relative to the build-object adhering to prescribed deposition and system constraints,
8. Generate the build-object and tool-head trajectories adhering to deposition and system constraints, and
9. Build the part.

The distinction between a "path" and "trajectory" in these steps highlights the distinction between positions and orientations purely in space (i.e., oriented points along a path) and positions and orientations evolving in time (i.e., oriented points along a trajectory). This distinction is critical for deposition methods as time dependence is dictated by the details of deposition (e.g., material feed rate and thermal considerations).

Execution of this new AM process requires, at a minimum, articulation of the build-object relative to the AM tool-head used to deposit material. Depending on build-object geometries, this procedure may further benefit from additional articulation of the AM tool-head to sufficiently reach and deposit material per the model specification. The fundamental dual manipulator concept explored in this work is shown in Figure 1. In this concept, industrial-style manipulators are used in coordination to move both the build-object and AM tool-head. This redundant approach expands the manipulation space of the system by providing several benefits including compensation for potential interference issues between the build-object and AM tool-head.

In this context, the combined system contains ≥ 12 DOF to command the coordinated trajectories of the manipulators (assuming each industrial manipulator contains six or more joints). The mapping that relates the relative trajectory for material deposition (prescribed in 6-DOF) to the coordinated trajectories of the manipulators is underdetermined. This provides flexibility as there may exist multiple (potentially infinite) sets of coordinated manipulator trajectories that produce the same relative trajectory for material deposition. This flexibility enables coordinated trajectories to be selected to eliminate interferences/collisions, enable the application of additional constraints on deposition (e.g., reorienting the part to reduce/eliminate support structure), etc.

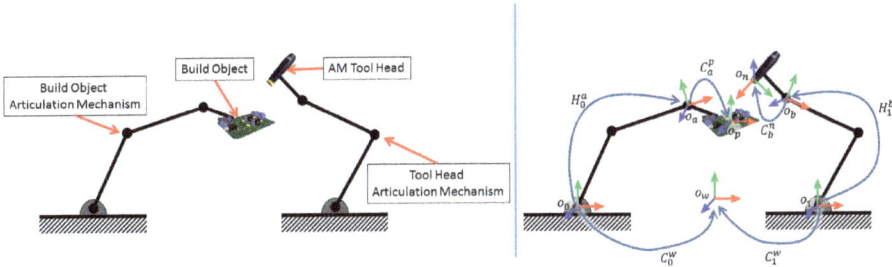

Figure 1. Illustration of fully articulated build-object and tool-head (**left**), and example coordinate frame assignments with associated transformation definitions (**right**) highlighting the minimum position/orientation measurements required to achieve conformal AM with this approach.

This paper presents the fundamental methods required to demonstrate conformal additive manufacturing. The methods presented include (1) the derivation of deposition paths given a prescribed set of layers; (2) the design, characterization, and control of a proof-of-concept testbed; and (3) the derivation and application of time evolving trajectories subject to the material deposition constraints and mechanical constraints of the testbed. Derivations are presented in a general context with examples extending beyond the proposed testbed. Results show the feasibility of conformal material deposition (i.e., onto and around existing surfaces) with multilayer, thick features.

Section 2 provides a summary of layering methods as applied to this approach using the prior work of Davis et al. [5]. Section 3 presents a new approach to defining paths for filling layers leveraging a projected ring approach as opposed to more common methods inspired by space filling curves [12]. Section 4 derives a general method to create coordinated deposition trajectories assuming manipulation of both the build-object and tool-head. Trajectory constraints are prescribed in the context of physical system limitations (e.g., velocity and acceleration constraints) and a simplified set of deposition constraints common to FDM methods. Section 5 reviews the system testbed design, associated geometric constraints, and coordinated control and provides a specific application example of deposition trajectory derivation based on the constraints of the testbed. Section 6 reviews the calibration of the system and summarizes experimental tests.

2. Review of Layering Methods

Davis et al. [5] present two approaches for defining conformal layers between two co-registered surfaces (the build-object and desired object geometry). The first method involves the use of a variable offset curve $\vec{x}_1(t;r)$ resulting from a parametrized curve $\vec{x}_0(t)$ as defined

$$\vec{x}_1(t;r) = \vec{x}_0(t) + r(t)\hat{N}(t). \tag{1}$$

Here, $r(t) \in \mathbb{R}^+$ is a parametrically-varying scalar and $\hat{N}(t)$ is the unit normal to $\vec{x}_0(t)$. For this method to be applicable, [5] assumes (1) the build-object is a convex geometry; (2) the desired object geometry is at most star-convex; and (3) the build object centroid is positioned relative to the desired part centroid to ensure intersections of the unit normal.

For the 2D case, two curves are given, γ_0 and γ_1 where $\gamma_0 \subset \gamma_1$. Here, γ_0 represents the 2D surface of the build object, and γ_1 represents the 2D surface of the desired part. For example, Figure 2 (left) illustrates a circular curve γ_0 and elliptic γ_1. For generality, curves are represented by piecewise parametric cubic splines fit to two sets of ordered points. Normals to γ_0 are defined at regular intervals, and intersections between these normals and γ_1 are calculated. Normal segments are then length parametrized, and variable offset curves can be defined using points along the parametrized normals.

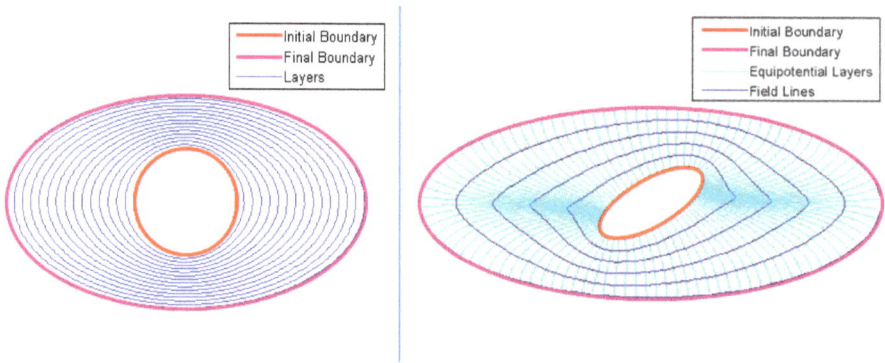

Figure 2. Variable offset curves derived between a build-object defined by a circle and desired object geometry defined by an ellipse (**left**); and Equipotential curves derived from solutions to Equation (2) (**right**).

Extending the concept of a variable offset curve into 3D to define variable offset surfaces is accomplished in a similar fashion to the approach taken for 2D. While the offset curves (or layers) produced appear evenly distributed, the algorithm is only applicable to a small subset of shapes.

To address non-convex geometries, [5] provide an alternate approach leveraging solutions to Laplace's equation for defining layers between curves. For an electrostatic potential field defined by $\varphi(x, y, z)$, Laplace's equation is given by

$$\triangledown^2 \varphi = \left(\frac{\partial^2}{\partial x^2} + \frac{\partial^2}{\partial y^2} + \frac{\partial^2}{\partial z^2} \right) \varphi(x, y, z) = 0 \tag{2}$$

and solved by applying boundary conditions. For this application, boundary conditions are defined by the surfaces of the build-object and desired part, and layers are defined as equipotential surfaces in the simulated electric field around the build-object. Figure 2 (right) illustrates a simple 2D example.

This method is well suited for arbitrary geometries and can be extended from simple shape examples in 2D to application relevant 3D geometries. Figure 3 shows the application of the Laplace approach to derive layers between a build-object defined as a populated circuit board, and a desired surface defined as a computer mouse.

Figure 3. Build-object defined as a 3D CAD model of a populated circuit board (left) and desired surface defined as 3D CAD model of a computer mouse. A cross-section of equipotential curves derived from solutions to (2) equation between the build-object and desired surface is shown in the center.

Both methods are further expanded to incorporate hollow-features (subject to geometric limitations) into the layering geometry. This allows layers to be defined both around the build-object and specified voids defined for weight reduction, material savings, etc. These layering methods will serve as the foundation for the approach presented in this paper.

It is of note that limitations still must be addressed before these methods can be generally applied; several of which are described in [5]. In addition to the limitations discussed in [5], issues of of layer "smoothness" and the uniformity of layer spacing are currently unaddressed. As can be seen in Figure 2 (right), areas associated with dense field lines produce layers with apparent protrusions that diminish closer to the exterior surface. While these layers are spaced "appropriately" under the constraints of [5], the layer geometry may prove difficult to realize in the deposition process. Similarly, inspection of Figures 2 and 3 suggests that the spacing between layers is not uniform at all points along a given layer surface. This implies that a single deposition pass may not sufficiently fill the space between the layers defined using the methods of [5].

3. Deposition Path Generation

Path or "scanning path" generation for existing AM systems is typically based on one or more space filling curve(s) where each discrete, flat layer is decomposed (typically relative to the outer surface of the part) and filled with material subject to prescribed infill constraints [12,13]. This approach has been extensively explored and applied across a wide variety of AM processes. For non-planar layers as proposed in this work, extensions of planar space filling curve approaches to non-planar layers is certainly feasible. As an example, any surface can be decomposed into a discrete set of open surfaces (e.g., cubed-sphere [14]), each discrete surface can then be mapped to a plane, and a desired space filling curve can be applied. Based on the "flatness" of each discrete surface, some additional steps may be necessary to maintain appropriate spacing between paths when mapped back to the non-planar surface.

For the FDM-based testbed considered in this work, we will explore an alternative approach inspired by the potential for layers defined by closed surfaces where it may be desirable to minimize unnecessary seams. Seams, in the context of FDM, are locations where a "material extrusion tool-path starts and ends on each closed part curve [15]". In the context of existing FDM processes, seams are only considered on the outer surface of the part and should typically occur once per layer. Using this definition, seams are effectively concealed by defining the start and end positions of the outer path within the part [16].

In the context of conformal AM, we will consider the total number of seams for every layer wherein a seam is defined as a point where a new deposition path begins or ends. Using a decomposition approach discussed above, the level of surface discretization will be proportional to one half of the total number of seams (assuming each discrete surface includes a tool-path start and end position that do not coincide). As a result, decomposing each layer into a discrete set of open surfaces provides a suboptimal solution. As an alternative, we consider an approach that, under ideal conditions, provides one contiguous path for layer geometries.

For the purposes of demonstration and without loss of generality, consider an ellipsoid defined parametrically about a body-fixed coordinate frame located at the centroid and aligned with the principal axes

$$
\begin{aligned}
x_1(u,v) &= c_1 cos(u) sin(v) \\
x_2(u,v) &= c_2 sin(u) sin(v) \\
x_3(v) &= c_3 cos(v).
\end{aligned}
\tag{3}
$$

Here, c_1, c_2, and c_3 define the magnitude of the principal axes; x_1, x_2, and x_3 define coordinates referenced to a body-fixed frame aligned with the principal axes (defined $\hat{x}_1$, $\hat{x}_2$, and $\hat{x}_3$) and located at the volumetric center; u is constrained to $u \in [0, 2\pi)$; and v is constrained to $v \in [0, \pi]$ as shown in Figure 4. To define paths, we consider a series of concentric "rings" offset along any given principal axis $\hat{x}_k \ \forall \ k \in \{1,2,3\}$. Use of the principal axes to define rings is independent of the parametrization. This provides a method suitable for any smooth, C^1 continuous surface. Without loss of generality, we assume that $k = 3$ given the parametrization provided in Equation (3). As a result, the ring for a given $x_3(v_i) = c_3 cos(v_i)$ is defined using

$$
\begin{aligned}
x_1(u, v_i) &= c_1 \cos(u) sin(v_i) \\
x_2(u, v_i) &= c_2 \sin(u) sin(v_i),
\end{aligned}
\tag{4}
$$

where $i \in \{0, 1, 2, \ldots, i_{max}\}$ denotes the discrete ring, $u \in [0, 2\pi)$, $v_i \subset [0, \pi]$, and v_0 can be assumed to be zero.

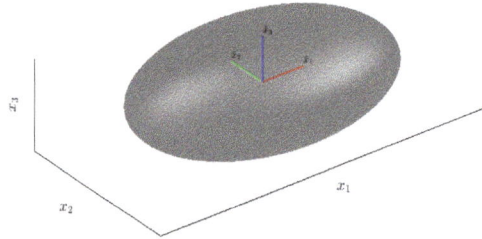

Figure 4. Ellipsoid with body-fixed coordinate frame located at the centroid and aligned with the principal axes $\hat{x}_1$, $\hat{x}_2$, and $\hat{x}_3$.

Spacing between concentric rings is defined by the effective width of deposited material. Assuming material is deposited with a fixed circular cross section of radius r, v_i is calculated based on the previous ring v_{i-1} subject to the constraint

$$\left| (x_1(u, v_i) - x_1(u, v_{i-1}), x_2(u, v_i) - x_2(u, v_{i-1}), x_3(u, v_i) - x_3(u, v_{i-1}))^{\mathsf{T}} \right| = 2r. \tag{5}$$

As is expected for all but the special case where $c_1 = c_2$, this approach yields a problematic result as $i \to i_{max}$ where trajectory begins to self-intersect. Results for the $c_1 = c_2$ case are presented in Figure 5 and general results are shown in Figure 6.

The condition of a self-intersecting trajectory for a given ring i is described by

$$\left| \left((x_1(u, v_i), x_2(u, v_i), x_3(u, v_i))^{\mathsf{T}} \Big|_{u = u_{i,p}^*} - (x_1(u, v_i), x_2(u, v_i), x_3(u, v_i))^{\mathsf{T}} \right) \right| = 2r. \tag{6}$$

If and when this condition occurs, the ideal solution of two seams per layer must be relaxed. In the case shown in Figure 5, no self intersection for any discrete ring occurs. As such, this layer is associated with a single start point, and a single end point. In Figure 5, we see a self intersection occur near $x_3 = -c_3$. In this case, a single path must be split or branched based on the p points of intersection $u_{i,p}^*$, where $u_{i,p}^*$ is defined using the condition presented in Equation (6). The resultant branched paths are described by

$$\begin{aligned} x_1(u_{i_j,j}, v_{i_j,j}) &= c_1 \cos(u_{i_j,j}) \sin(v_{i_j,j}) \\ x_2(u_{i_j,j}, v_{i_j,j}) &= c_2 \sin(u_{i_j,j}) \sin(v_{i_j,j}). \end{aligned} \tag{7}$$

In the case of Figure 6, a single intersection occurs, and the subsequent split paths are defined by first cropping the remaining two surfaces. Once cropped, the two surfaces are filled to define the branches using steps matching those of the original decomposition process with the only exception being the definition of the first ring. In this case, $x_1(u_{0,j}, v_{0,j})$, $x_2(u_{0,j}, v_{0,j})$, and $x_3(u_{0,j}, v_{0,j})$ are defined along the cropped edge of the surface. This basic procedure of defining the initial ring along an edge further applies to discretizing open surfaces.

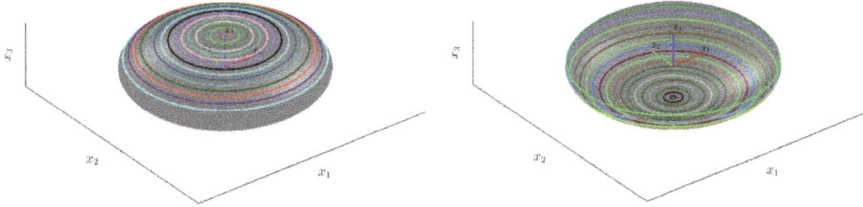

Figure 5. Ellipsoid where $c_1 = c_2$ with overlaid concentric paths propagated from $x_3 = c_3$. Paths on the top of the ellipsoid are shown on the left, and paths on the bottom are shown on the right.

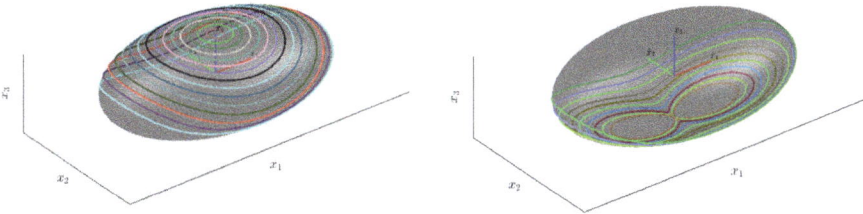

Figure 6. Ellipsoid where $c_1 \neq c_2$ with overlaid concentric paths propagated from $x_3 = c_3$. Paths on the top of the ellipsoid are shown on the left, and paths on the bottom are shown on the right.

With the entire layer decomposed into rings, one or more paths can be created to cover the surface. To do so, we first parametrize the system over $w \in [0, 1]$, defining the parametrized coordinates of ring i using the vector $\vec{x}_i(w)$ defined in Equation (8).

$$\vec{x}_i(w) = \begin{pmatrix} x_1(u, v_i; w) \\ x_2(u, v_i; w) \\ x_3(u, v_i; w) \end{pmatrix} \tag{8}$$

Using this parametrization, rings are "cut" subject to the deposition constraint proposed in Equation (5) such that

$$|\vec{x}_i(0) - \vec{x}_i(1)| = 2r. \tag{9}$$

Doing so further restricts the bounds on u on a ring-by-ring basis. We define u_i for each cut ring such that $u_i(w) \subset [0, 2\pi)$. To maintain aligned cuts, selection of $u_i(0)$ for $i > 0$ is defined such that

$$|\vec{x}_i(0) - \vec{x}_{i-1}(0)| = |\vec{x}_i(1) - \vec{x}_{i-1}(1)|. \tag{10}$$

Once cut, rings must be smoothly connected or "stitched" to create a contiguous path. Assuming small r, stitches are defined by refining the end conditions of each ring. To do so, we introduce offset conditions for w for each ring and end point such that

$$|\vec{x}_i(\Delta w_i(0)) - \vec{x}_i(0)| = r \tag{11}$$

$$|\vec{x}_i(\Delta w_i(1)) - \vec{x}_i(1)| = r \tag{12}$$

where $\Delta w_i(0)$ defines an offset from $\vec{x}_i(0)$, and $\Delta w_i(1)$ defines an offset from $\vec{x}_i(1)$. For small r, we note that Euclidean norm is approximately equal to the distance along the surface. This allows us to define a stitch between each ring using an arc of constant curvature (approximately equal to r) from $\vec{x}_{i-1}(\Delta w_{i-1}(0))$ to $\vec{x}_i(\Delta w_i(0))$ and from $\vec{x}_{i-1}(\Delta w_{i-1}(1))$ to $\vec{x}_i(\Delta w_i(1))$ for all $i \in \{1, 2, \ldots, i_{max}\}$.

Once rings are stitched, a wave function with bounds at 0 and 1 defined over $\xi \in [0, i_{max} + 1)$ (e.g., Equation (13)) is used to define a contiguous path.

$$w(\xi) = \frac{1}{2} \sin\left(\pi\xi - \frac{\pi}{2}\right) + \frac{1}{2} \tag{13}$$

Defining $w(\xi)$ per Equation (13) allows the deposition path to be defined

$$\vec{x}(\xi) = \vec{x}_{\lfloor \xi \rfloor}(w; \xi). \tag{14}$$

For this application, paths are extended from a three dimensional position, to a 5-DOF pose (position in three dimensions and deposition direction prescribed by two angles). This is critical when prescribing tool orientation during deposition. Intuitively, deposition must occur with the tool-head aligned with the surface normal. Using the parametrization provided in Equation (3), the surface normal is defined

$$\vec{N} = \frac{\partial \vec{x}}{\partial u} \times \frac{\partial \vec{x}}{\partial v}, \tag{15}$$

where $\vec{x} = (x_1(u,v), x_2(u,v), x_3(v))^{\mathsf{T}}$ and $\vec{N}$ defines the surface normal relative to the body-fixed coordinate frame. With a deposition path and orientation prescribed, the deposition trajectory can be defined.

4. Deposition Trajectory Generation

For this FDM inspired application, trajectories are defined by parameterizing paths in time subject to the bounds of the deposition tool-head. Assume the tool-head extrudes material at a linear rate d which is continuously variable between 0 (no material is deposited) and d_{max} (the maximum allowable rate of deposition). At any given point along the trajectory, the instantaneous speed must be bounded by d. Equation (16) defines this relationship between speed and deposition rate where $\vec{T}(t)$ defines the time-evolving tangent to the path and instantaneous speed is defined as the Euclidean norm of the tangent.

$$\left|\vec{T}(t)\right| \leq d_{max} \tag{16}$$

Parameterizing the path with respect to arc length simplifies the derivation of the trajectory noting that, by definition

$$\left|\vec{T}(s)\right| = \left|\frac{d\vec{x}(s)}{ds}\right| = 1 \tag{17}$$

where $s \in [0, 1]$. This enables a constant deposition rate of $\mathbf{d}^*$ defined within the bounds of d to be applied. Given the path parametrized by arc length, the function $s(t)$ can be defined noting Equation (18).

$$\left|\vec{T}(t)\right| = \left|\frac{\partial \vec{x}(s)}{\partial s}\frac{\partial s(t)}{\partial t}\right| = \mathbf{d}^*. \tag{18}$$

Noting that $\frac{\partial s(t)}{\partial t}$ is a scalar and $s(t)$ is strictly increasing, $s(t)$ can be defined

$$s(t) = \int_0^t \mathbf{d}^* d\tau = \mathbf{d}^* t. \tag{19}$$

Build-object and tool-head trajectories are derived from the deposition trajectory using a model-based approach. Assuming the redundancy present in the proposed dual manipulator system, there is the potential for multiple sets of coordinated manipulator trajectories that result in the same deposition trajectory. This allows candidate sets to be evaluated in simulation to check for issues related to interference, collision, and joint velocities/accelerations exceeding the physical capabilities of the hardware. This redundancy also makes it possible to impose additional constraints on the system to adhere to desired criteria (e.g., actively reorienting the deposition direction relative to gravity).

The primary drawback to this level of redundancy is the extensive search space associated with two coordinated manipulators. For the purposes of this work, the search space is reduced by assuming a trajectory for the build-object, and deriving an interference and collision-free trajectory for the tool-head. This is accomplished using a variety of available tools (e.g., MoveIt! [17]). While effective for this application, this approach may be improved using methods from existing research [18,19], however this is outside of the scope of this work.

5. Testbed Design

5.1. Hardware Overview

An asymmetric set of two independent six degree-of-freedom (6-DOF) manipulators (UR5 and UR10, Universal Robots A/S, Odense, Denmark), a single gripper (2-Finger Adaptive Robot Gripper, Robotiq Inc., Lévis, QC, Canada), a single tool-head (3Doodler v1.0, WobbleWorks, LLC., Somerville, MA, USA), and a 14 camera motion capture system (OptiTrack Prime 41, NaturalPoint Inc., Corvallis, OR, USA) comprise the system testbed. The asymmetry in manipulator geometry enables a large shared workspace in $x, y, z,$ and $\hat{z}$ tool coordinates. Here, $x, y,$ and z denote the tool position, and $\hat{z}$ denotes the z-direction of the tool frame. Figure 7 provides an annotated view of the system.

Control of the testbed consists of four key items (1) Interfacing and controlling the Universal Robot hardware to execute a coordinated set of smooth, prescribed trajectories; (2) Interfacing the Robotiq gripper to reliably respond to a known command set; (3) Creating an electronic interface with the 3Doodler capable of responding to a known command set to control material feed rate; and (4) Registering and tracking the build object and tool-head using the motion capture system. The following sections will address items (1) and (4) in detail.

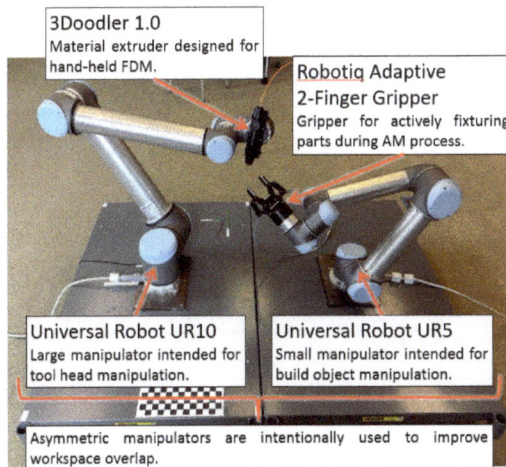

Figure 7. System testbed highlighting key components.

5.2. Controller Design

Interfacing and controlling the Universal Robot hardware to execute a coordinated set of smooth, prescribed trajectories requires the development of an on-board intermediate control algorithm. A script implementing an intermediate controller was developed to run directly on the UR operating system. This script leverages a modified PID approach wherein desired discrete set of joint positions ($\vec{q}_d(t_i)$) and velocities ($\dot{\vec{q}}_d(t_i)$) are sent to the manipulator, and the controller generates a continuous commanded joint velocity ($\dot{\vec{q}}(t)$) for the manipulator. The current implementation utilizes a proportional controller, $k_p > 0$, where discrete time steps are denoted using t_i, and continuous terms (available as direct feedback on-board the UR operating system) are denoted as functions of t.

$$\dot{\vec{q}}(t) = k_p \left(\vec{q}_d(t_i) - \vec{q}(t) \right) \tag{20}$$

To interface each Universal Robot, a MATLAB class [20] was created wrapping existing functionality from the available URX Python Library [21]. This class develops a custom command structure for sending and receiving information to/from the control script. This allows the UR to respond to a continuous stream of joint position/velocity waypoints in a smooth manner. Commands are sent to each manipulator via a TCP/IP connection from a PC running a single instance of MATLAB. Time-stamped waypoints are calculated off-line and sent to each manipulator at a known interval. This enables coordination driven by the clock of the host PC.

5.3. System Kinematics

As was introduced in Figure 1, successful operation of this dual manipulator testbed requires the measurement and estimation of numerous transformations between coordinate frames in space. Frames are initially assigned using available measurements from the motion capture system and joint measurements from each Universal Robot. The motion capture system provides position and orientation (also referred to as pose) measurements of rigid configurations of reflective markers relative to a static world frame. Each Universal Robot provides the pose of their end-effector relative to their respective base frame. This introduces Frame w, Frame b_1, Frame b_2, Frame E_1, and Frame E_2 defined in Table 2; where the UR5 provides $H_{E_1}^{b_1}$ (the pose, represented as a rigid body transformation, of Frame E_1 *relative* to Frame b_1), and the UR10 provides $H_{E_2}^{b_2}$.

Table 2. Coordinate frame definitions for the dual manipulator testbed.

Label	Description
Frame W	Motion Capture World Frame
Frame b_1	UR5 Base Frame
Frame b_2	UR10 Base Frame
Frame E_1	UR5 End-effector Frame
Frame E_2	UR10 End-effector Frame
Frame m_1	Marker Frame rigidly fixed relative to Frame b_1
Frame m_2	Marker Frame rigidly fixed relative to Frame b_2
Frame T_1	The body-fixed coordinate frame of the build-object (rigidly fixed relative to Frame E_1)
Frame T_2	The tool-head coordinate frame with $\hat{z}$ aligned with the material feed direction and offset from the nozzle per manufacturer recommendations (rigidly fixed relative to Frame E_1)

Noting that no pose information between the manipulators is known, we fix a rigid set of reflective markers relative to the base of each manipulator. This introduces Frame m_1 and Frame m_2 (Table 2) where the motion capture provides $H_{m_1}^W$ and $H_{m_2}^W$. Section 5.4 addresses the experimental estimation of $\mathbf{H}_{m_1}^{b_1}$ and $\mathbf{H}_{m_2}^{b_2}$.

To account for the build-object and tool-head, we introduce Frame T_1 and Frame T_2 (Table 2) where CAD models of the gripper and build-object provide an initial estimate of $\mathbf{H}_{T_1}^{E_1}$, and a CAD model of the tool-head provides an initial estimate of $\mathbf{H}_{T_2}^{E_2}$. Section 5.4 addresses refinement and validation of $\mathbf{H}_{T_1}^{E_1}$ and $\mathbf{H}_{T_2}^{E_2}$. A simulation of the testbed with labeled frame assignments is provided in Figure 8.

These transformations are combined to provide $H_{T_2}^{T_1}$, the rigid body transformation relating the tool-head to the body-fixed frame of the build-object using

$$H_{T_2}^{T_1} = \mathbf{H}_{E_1}^{T_1} H_{b_1}^{E_1} \mathbf{H}_{m_1}^{b_1} H_W^{m_1} H_{m_2}^W \mathbf{H}_{b_2}^{m_2} H_{E_2}^{b_2} \mathbf{H}_{T_2}^{E_2}. \tag{21}$$

Noting the derivations in Sections 3 and 4, Frame T_1 is analogous to the body-fixed frame of each surface, path, and trajectory. Therefore, the evolution of $H_{T_2}^{T_1}$ with time is directly prescribed by the trajectory.

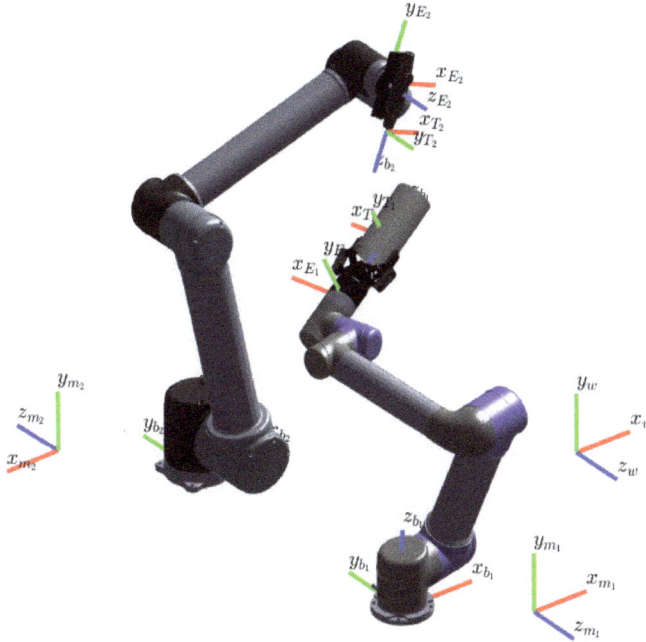

Figure 8. Simulation of system testbed highlighting frame definitions.

5.4. Calibration

System calibration is performed by first creating calibration rigid bodies for each manipulator. Noting the manufacturer's assignment of Frame E_1 for the UR5, and Frame E_2 for the UR10, two calibration objects are designed to precisely place a set of reflective markers at known locations relative to each. A rendering of the UR5 rigid body is shown in Figure 9, and the fabricated rigid bodies for both the UR5 and UR10 are shown in Figure 10.

Using [22], $H_{E_1}^W$ and $H_{E_2}^W$ are estimated from the measured marker locations returned by the motion capture system. Using these measurements, $\mathbf{H}_{m_1}^{b_1}$ and $\mathbf{H}_{m_2}^{b_2}$ are estimated

$$\mathbf{H}_{m_i}^{b_i} = H_{E_i}^{b_i} H_W^{E_i} H_{m_i}^{W} \ \forall i \in \{1, 2\} \tag{22}$$

where $H_{E_i}^{b_i}$ is measured by the respective Universal Robot, $H_W^{E_i}$ and $H_{m_i}^{W}$ are measured using the motion capture, and $H_W^{E_i} = \left(H_{E_i}^W\right)^{-1}$. Given the inherent uncertainty associated with experimentally measured parameters, we refine the estimate of $\mathbf{H}_{m_i}^{b_i}$ for each manipulator by collecting a large number of samples over the manipulator workspace and calculating the mean of the resultant set of $\mathbf{H}_{m_i}^{b_i}$ using [23].

Estimates of $\mathbf{H}_{T_1}^{E_1}$ and $\mathbf{H}_{T_2}^{E_2}$ can be refined using techniques commonly applied to computer assisted surgical systems and computer vision. Using a precision machined probe [24] with known correspondence between the tip position and body-fixed frame, points along the outer surface of the build-object can be digitized relative to Frame W. Using these points and associated CAD models of the build-object and tool-head, $\mathbf{H}_{T_1}^{E_1}$ and $\mathbf{H}_{T_2}^{E_2}$ can be refined using [25]. Further refinement of $\mathbf{H}_{T_2}^{E_2}$ can be performed by precisely estimating nozzle tip position using a pivot calibration [26].

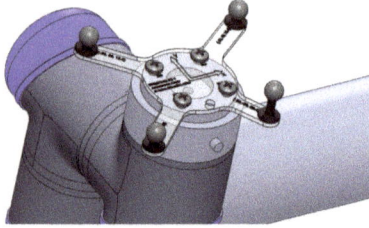

Figure 9. CAD model of the calibration rigid body designed for the UR5 manipulator.

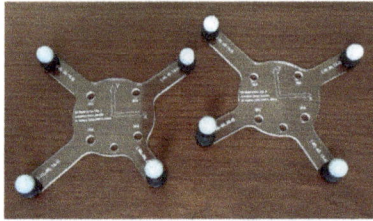

Figure 10. Fabricated calibration rigid bodies for the UR5 (**left**) and UR10 (**right**) manipulators.

6. Results

Experimental validation was conducted by evaluating the techniques discussed in this work applied to a single layer of deposition onto a 75 mm long cylindrical build object matching the outside diameter of standard 3 in (76.2 mm) schedule 40 PVC pipe with an outside diameter of 88.9 mm. Ring spacing was defined assuming that the deposition radius of r is equal to the 1.0 mm (one half of the extrusion nozzle diameter increased by a margin of 0.5 mm). Note that, in practice, the value for r should be defined experimentally and is typically larger than the extrusion nozzle diameter. A deposition path was derived using the methods described in Section 3, and a trajectory was derived using the methods described in Section 4 with $\mathbf{d}^* = 40$ mm/s. Figure 11 (left) shows the cylindrical layer decomposed into cut rings, and Figure 11 (right) shows the stitched rings used to generate the path and trajectory.

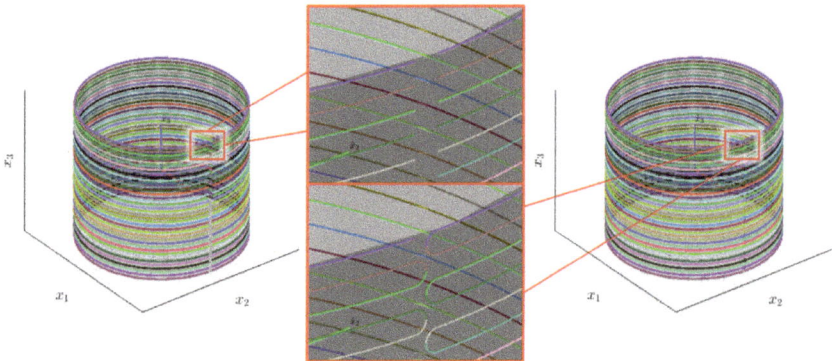

Figure 11. Cylindrical surface decomposed into cut rings (**left**) and stitched rings used to generate the deposition path (**right**).

6.1. Calibration Results

System calibration was conducted using the calibration rigid bodies described in Section 5.4. A total of 28 samples were taken from each manipulator over a discrete set of joint configurations defined over the outside of the workspace. For each arm, the mean transformation relating Frame m_i to b_i was calculated. Calibration results are analyzed using the RMS error between the fixed marker locations on the base of each manipulator measured by the motion capture, and the marker locations estimated using $H_{m_i}^W$ defined

$$H_{m_i}^W = H_{E_i}^W H_{b_i}^{E_i} \mathbf{H}_{m_i}^{b_i}. \tag{23}$$

Results from the UR5 calibration are presented in Figure 12 (left), and results from the UR10 calibration are presented in Figure 12 (right).

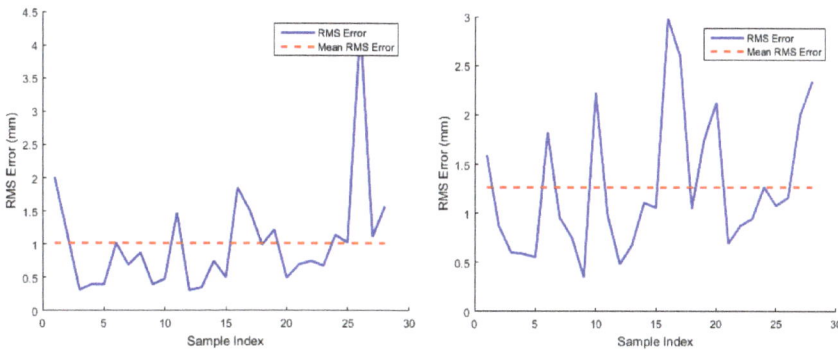

Figure 12. RMS error calculated between measured marker locations and marker location estimates calculated with $\mathbf{H}_{m_1}^{b_1}$ (**left**) and $\mathbf{H}_{m_2}^{b_2}$ (**right**).

6.2. Controller Results

The intermediate controller was analyzed by comparing the control signal sent to each robot to the actual end-effector position while executing the deposition trajectory. Results from the UR5 are presented in Figure 13 (left), and results from the UR10 are presented in Figure 13 (right).

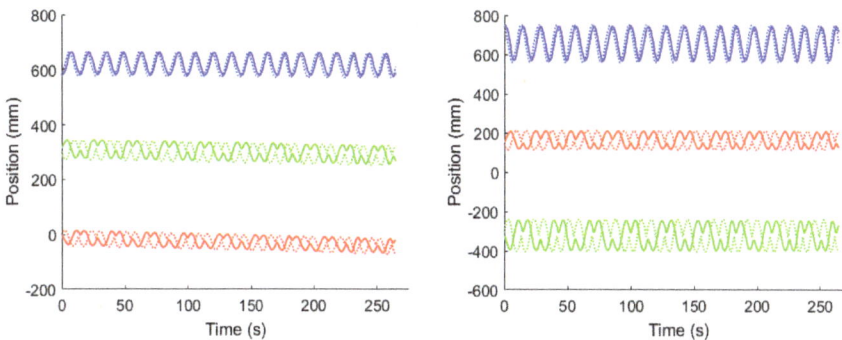

Figure 13. Comparison between actual end-effector position and commanded end-effector position for the UR5 (**left**) and UR10 (**right**). Actual position is represented using a solid line, and command position is represented using a dashed line. x, y, and z positions are differentiated using red, green, and blue respectively.

6.3. System Performance

The overall performance of the system was analyzed by comparing the commanded and measured deposition trajectories estimated using $H_{T_2}^{T_1}$. Comparison results for the position in the body-fixed x-direction are presented in Figure 14, the body-fixed y-direction in Figure 15, and the body-fixed z-direction in Figure 16.

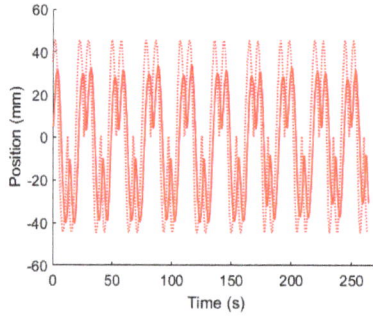

Figure 14. Comparison between the x-position of the actual deposition trajectory and the commanded deposition trajectory. Actual position is represented using a solid line, and command position is represented using a dashed line.

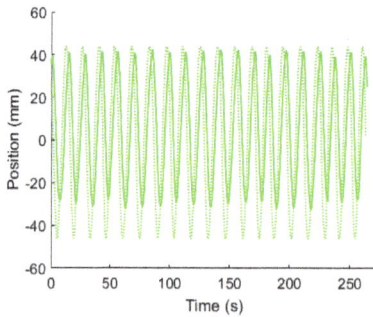

Figure 15. Comparison between the y-position of the actual deposition trajectory and the commanded deposition trajectory. Actual position is represented using a solid line, and command position is represented using a dashed line.

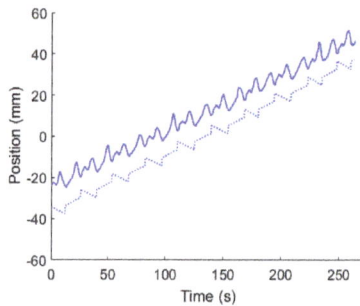

Figure 16. Comparison between the z-position of the actual deposition trajectory and the commanded deposition trajectory. Actual position is represented using a solid line, and command position is represented using a dashed line.

7. Discussion

System calibration was quantified by calculating mean RMS error associated with the difference between measured and estimated marker positions. This RMS error is a measure of accuracy for $\mathbf{H}_{m_i}^{b_i}$ that accounts for errors in both position and orientation associated with the estimated transformation. Experiments showed an RMS error for $\mathbf{H}_{m_1}^{b_1}$ of 1.01 mm, and an RMS error for $\mathbf{H}_{m_2}^{b_2}$ of 1.27 mm. These errors are reasonable when considering the published worst case performance specifications from both Univeral Robots and NaturalPoint; however this error must be reduced before actual material deposition can occur. Error reduction using the current and proposed system tools discussed in Section 5.4 is the subject of ongoing work. An additional solution for error reduction using a shared, precision machined base for both manipulators is under consideration, however this may limit system versatility.

Experiments showed that, for the cyclic trajectories used in this paper, the end-effector tracking errors reached a steady state amplitude of approximately 20 mm and phase lag of π radians. However, both manipulators were found to have the same tracking error characteristics, so the relative trajectories remained within spatial tolerances of the tool-head. Iteration of the control design including integral and derivative terms to improve tracking performance is the subject of ongoing work.

Comparisons of the commanded and measured trajectories show discrepancies that reach and exceed 10 mm in position data, however these results are expected given the analysis of system calibration and control. Given the RMS error measurements associated with $\mathbf{H}_{m_1}^{b_1}$ of 1.01 mm, and an RMS error for $\mathbf{H}_{m_2}^{b_2}$ of 1.27 mm; these discrepancies are explained primarily by an orientation misalignment associated with the estimated transformations relating the marker frame to the base frame for each robot. Refining system calibration using the methods described in Section 5.4 will further reduce this error. Analysis of the published performance capabilities for the UR manipulators and motion capture system suggest that this error can be reduced to better than ± 1.0 mm. Performance can be further improved using a shared, precision machined base for both manipulators; eliminating tracking errors associated with the motion capture system. As mentioned previously, this approach may limit system versatility. Methods to reduce this error using the tools discussed in Section 5.4 are the subject of ongoing work.

Additional analysis of the comparisons of the commanded and measured trajectory results also suggest that the selection of $\mathbf{d}^*$ may have an effect on deposition tracking performance. While the prescribed trajectory was within the performance limitations of the system, a reduced value of $\mathbf{d}^*$ may provide improved tracking accuracy. Further investigation into the relationship between the selection of $\mathbf{d}^*$ and tracking performance is the subject of ongoing work.

8. Conclusions

We have presented a systematic approach for multilayer conformal additive manufacturing inspired by the work presented in [5]. This work included a new approach to generate appropriately spaced paths to fill the surface of three dimensional layers. Paths were used to derive time evolving trajectories using arc length parametrization and the extrusion rate of the tool-head. The application of these methods to a relevant hardware system was discussed, and topics including control, characterization, and calibration were addressed. Results from this work suggest the feasibility of this approach in a relevant context; however future work is required to refine system performance.

Acknowledgments: This work was supported by Office of Naval Research FY16 Grant No. N0001416WX00796. The authors thank Greg Chirikjian, Josh Davis, and Jin Seob Kim for their contributions to the foundation of this work, and ENS Kevin Strotz for his contributions to the Universal Robots MATLAB Toolbox as a TAD Ensign during the summer of 2016.

Author Contributions: Kutzer M.D.M. developed path and trajectory generation methods. DeVries L.D. defined and evaluated intermediate controller for the manipulator. Kutzer M.D.M. designed experiments; Kutzer M.D.M. and DeVries L.D. performed experiments and analyzed data.

Conflicts of Interest: The authors declare no conflict of interest.

1. Scheck, C. Introduction to Additive Manufacturing (AM 101). Available online: http://www.nsrp.org/wp-content/uploads/2016/01/Presentation-Joint_120815-AM_101-NSWCCD.pdf (accessed on 19 May 2016).
2. Frazier, W.E. Direct digital manufacturing of metallic components: Vision and roadmap. In Proceedings of the 21st Annual International Solid Freeform Fabrication Symposium, Austin, TX, USA, 9–11 August 2010; pp. 717–732.
3. Goehrke, S. NAVAIR Has Big Plans for Additive Manufacturing. Available online: https://3dprint.com/58698/navair-additive-manufacturing/ (accessed on 25 May 2016).
4. Harper, J. Military 3D Printing Projects Face Challenges. *Natl. Def. Ind. Assoc. Bus. Technol. Mag.* **2015**, *100*, 24.
5. Davis, J.D.; Kutzer, M.D.; Chirikjian, G.S. Algorithms for Multilayer Conformal Additive Manufacturing. *J. Comput. Inf. Sci. Eng.* **2016**, *16*, 021003.
6. Gibson, I.; Rosen, D.W.; Stucker, B. *Additive Manufacturing Technologies: Rapid Prototyping to Direct Digital Manufacturing*; Springer: New York, NY, USA, 2010.
7. Radtke, D.; Zeitner, U.D. Laser-lithography on non-planar surfaces. *Opt. Express* **2007**, *15*, 1167–1174.
8. Xie, Y.; Lu, Z.; Jingli, F.L.; Yongjun, Z.; Zhao, J.; Weng, Z. Lithographic fabrication of large diffractive optical elements on a concave lens surface. *Opt. Express* **2002**, *10*, 1043–1047.
9. Adams, J.J.; Duoss, E.B.; Malkowski, T.F.; Motala, M.J.; Ahn, B.Y.; Nuzzo, R.G.; Bernhard, J.T.; Lewis, J.A. Conformal Printing of Electrically Small Antennas on Three-Dimensional Surfaces. *Adv. Mater.* **2011**, *23*, 1335–1340.
10. Paulsen, J.; Renn, M.; Christenson, K.; Plourde, R. Printing conformal electronics on 3D structures with Aerosol Jet technology. In Proceedings of the Future of Instrumentation International Workshop (FIIW), Gatlinburg, TN, USA, 8–9 October 2012; pp. 1–4.
11. Directed Energy Deposition. Available online: http://www.lboro.ac.uk/research/amrg/about/the7categoriesofadditivemanufacturing/directedenergydeposition/ (accessed on 25 May 2016).
12. Bertoldi, M.; Yardimci, M.; Pistor, C.; Guceri, S. Domain decomposition and space filling curves in toolpath planning and generation. In Proceedings of the 1998 Solid Freeform Fabrication Symposium, Austin, TX, USA, 11–13 August 1998; pp. 267–274.
13. Yang, Y.; Fuh, J.Y.; Loh, H.T. An efficient scanning pattern for layered manufacturing processes. In Proceedings of the IEEE International Conference on Robotics and Automation (2001 ICRA), Seoul, Korea, 21–26 May 2001; Volume 2, pp. 1340–1345.
14. Sadourny, R. Conservative finite-difference approximations of the primitive equations on quasi-uniform spherical grids. *Mon. Weather Rev.* **1972**, *100*, 136–144.
15. Stratasys Ltd. Optimizing Seam Location. Available online: http://usglobalimages.stratasys.com/Main/Files/Best%20Practices_BP/BP_OptimizingSeamLocation.pdf?v=635817995301407599 (accessed on 25 May 2016).
16. Hopkins, P.E.; Holzwarth, D.J. Seam Concealment for Three-Dimensional Models. U.S. Patent 8,974,715, 10 March 2015.
17. Sucan, I.A.; Chitta, S. MoveIt! Available online: http://moveit.ros.org (accessed on 17 May 2017).
18. Latombe, J.C. *Robot Motion Planning*; Kluwer Academic Publishers: Norwell, MA, USA, 1991.
19. Ata, A.A. Optimal trajectory planning of manipulators: A review. *J. Eng. Sci. Technol.* **2007**, *2*, 32–54.
20. Kutzer, M. MATLAB Toolbox for UR Manipulators. Available online: https://www.usna.edu/Users/weapsys/kutzer/_Code-Development/UR_Toolbox.php (accessed on 22 May 2017).
21. Roulet-Dubonnet, O. Python Library to Control a Robot from "Universal Robots". Available online: https://github.com/SintefRaufossManufacturing/python-urx (accessed on 22 May 2017).
22. Eggert, D.W.; Lorusso, A.; Fisher, R.B. Estimating 3-D rigid body transformations: A comparison of four major algorithms. *Mach. Vis. Appl.* **1997**, *9*, 272–290.
23. Long, A.W.; Wolfe, K.C.; Mashner, M.J.; Chirikjian, G.S. The banana distribution is Gaussian: A localization study with exponential coordinates. In *Robotics: Science and Systems VIII*; MIT Press: Cambridge, MA, USA 2013; pp. 265–272.
24. Chen, E.; Sati, M.; Croitoru, H.; Tate, P.; Fu, L. Computer-Assisted Surgical Positioning Method and System. U.S. Patent 10/467,445, 6 May 2004.

25. Mitra, N.J.; Gelfand, N.; Pottmann, H.; Guibas, L. Registration of point cloud data from a geometric optimization perspective. In Proceedings of the 2004 Eurographics/ACM SIGGRAPH Symposium on Geometry Processing, Nice, France, 8–10 July 2004; pp. 22–31.

26. Otake, Y.; Armand, M.; Sadowsky, O.; Armiger, R.; Kutzer, M.; Mears, S.; Kazanzides, P.; Taylor, R. An image-guided femoroplasty system: Development and initial cadaver studies. *Proc. SPIE* **2010**, *7625*, 76250P.

technologies

MDPI

Article

Compression Tests of ABS Specimens for UAV Components Produced via the FDM Technique

Salvatore Brischetto *, Carlo Giovanni Ferro, Paolo Maggiore and Roberto Torre

Department of Mechanical and Aerospace Engineering, Politecnico di Torino, corso Duca degli Abruzzi 24, 10129 Torino, Italy; carlo.ferro@polito.it (C.G.F.); paolo.maggiore@polito.it (P.M.); robertotorre.mb@gmail.com (R.T.)
*   Correspondence: salvatore.brischetto@polito.it; Tel.: +39-011-090-6813

Academic Editor: Manoj Gupta
Received: 5 April 2017; Accepted: 3 May 2017; Published: 5 May 2017

Abstract: Additive manufacturing has introduced a great step in the manufacturing process of consumer goods. Fused Deposition Modeling (FDM) and in particular 3D printers for home desktop applications are employed in the construction of prototypes, models and in general in non-structural objects. The aim of this new work is to characterize this process in order to apply this technology in the construction of aeronautical structural parts when stresses are not excessive. An example is the construction of the PoliDrone UAV, a multicopter patented, designed and realized by researchers at Politecnico di Torino. For this purpose, a statistical characterization of the mechanical properties of ABS (Acrylonitrile Butadiene Styrene) specimens in compression tests is proposed in analogy with the past authors' work about the tensile characterization of ABS specimens. A desktop 3D printer, including ABS filaments as the material, has been employed. ASTM 625 has been considered as the reference normative. A capability analysis has also been used as a reference method to evaluate the boundaries of acceptance for both mechanical and dimensional performances. The statistical characterization and the capability analysis are here proposed in an extensive form in order to validate a general method that will be used for further tests in a wider context.

Keywords: Fused Deposition Modeling (FDM); 3D printing process; mechanical and dimensional properties; compression tests; Acrylonitrile Butadiene Styrene (ABS); statistical process control; compression modes

---

1. Introduction

Unmanned Aerial Vehicles (UAVs) or Remotely-Piloted Vehicles (RPVs) are a group of airplanes and helicopters that can fly autonomously or remotely controlled. Nowadays, most scientists and technicians are writing the rules to integrate in a yet crowded airspace the new subject, which potentially can cause severe damages to civilian airplanes [1,2] or can violate the privacy of people [3]. The Italian Authority for Civil Aviation regulates the capability of such aircraft in relation to the VLOS (Visual Line Of Sight) operations, and it limits the MTOW (Maximum Take Off Weight) to 25 kg (approximatively 50 lbs) [4]. The Civil Authority also defines the differences between critical operations (above people or in ATZ (Aerodrome Traffic Zone)) and no-critical operations (e.g., no-crowded areas). The market of so-called drones has had a enormous growth due to their simplicity of piloting, low cost and expansion of the applications from package delivering to farming or monitoring [5]. Nowadays, the most diffused type of RPVs is the multi-rotor, a sort of helicopter with three, four, six or eight arms which can hover above a place and can have vertical take-off and landing. Different applications (e.g., patrol, package delivering, aerial photography) usually require different types of multi-copters forcing the end-user to have in its disposal several machines. In order to reduce the cost of maintenance and purchase, the multipurpose and modular drone called PoliDrone has been patented as an innovative solution [6], which allows reconfiguring the machine in different ways changing the number of arms and propellers and the geometry of the vehicle.

In order to emphasize the diffusion of this platform, Fused Deposition Modeling (FDM) has been chosen as the construction technique to allow anyone with a 3D desktop printer to construct a prototype in a fast and economical way [7]. Using a low-cost building technology, it is possible to build each part of the flying machine at home reducing the spare parts' cost and the supply chain management; see the comparison with a drone factory such as Parrot or DJI. The model of part replacements and improvements on the main frame will be delivered as reported in [8]. A render of the prototype, which successfully flew for the first time on 4 July 2016, can be seen in Figure 1. PoliDrone is a multipurpose modular drone with adjustable arms produced via the FDM additive manufacturing process [9]. The combination of only eight basic constituent elements used in different ways and number allows 12 different configurations. These configurations are 3-, 4-, 6- and 8-arm configurations with the possibilities of single rotor, double rotor or system rotor + inflatable element per arm. This idea is completely new if compared with other modern drones proposed in [10–13].

Figure 1. Render of PoliDrone, a multipurpose and modular UAV.

The first PoliDrone prototype has been produced, via the 3D FDM printing process, using PLA (PolyLacticAcid) for all of the elements [9]. PLA is a green and recyclable material. It is quite easy to print. The main aim of the project is the construction of a second prototype with a total weight (including a pay load of 0.5 kg) less than 2 kg. In this way, we can take advantage of the facilities provided by the ENAC (Ente Nazionale Aviazione Civile or National Authority for Civil Aviation) regulation. In order to obtain this aim, three main steps must be followed. The first step is the definition of a new geometry for the prototype where some parts are redesigned. The second step is the use of a new material in combination with the FDM technique: ABS (Acrylonitrile Butadiene Styrene) in place of PLA. ABS is lighter than PLA (better specific properties), and it does not have deterioration of its properties through time (PLA has deterioration). However, the FDM printing process is not so easy if combined with the use of ABS [14,15]. The third step is the appropriate Finite Element (FE) analysis of the prototype and the relative structural optimization. In order to perform such an analysis, the ABS properties must be known. We know the properties of the ABS filament, but after the FDM process, such properties change and become unknown. We need these properties in order to perform a correct FE analysis. ABS has been chosen due to its good mechanical properties combined with a reduced weight. In order to design and optimize the primary structure of the multi-copter, knowing the applied loads, it is mandatory to have the material properties with a statistical level of confidence. This feature is necessary because during the extrusion process, there are several machine parameters that can influence the mechanical properties of the finished pieces [14]. Moreover, for a flying product, not only the mechanical properties are requested, but also the dimensional capabilities of the machine must be investigated in order to design the correct machine drawing, allowing perfect joints between different subparts [15].

It is important to notice that ABS can be polymerized in varying proportions and that the manufacturing process and machine precision can influence its properties. In general, three different directions $(1,2,3)$ of building are possible in the case of the FDM 3D printing process. For each building direction, the raster orientation can also be selected (e.g., $+45°/-45°$ in our study cases). For these reasons, a complete characterization is necessary for each building direction. In order to have a first satisfactory ABS characterization, a possible and general work plan could be:

- Tensile characterization for building Direction 1 for the specimen production
- Compressive characterization for building Direction 1 for the specimen production
- Bending characterization for building Direction 1 for the specimen production
- Tensile characterization for building Direction 2 for the specimen production
- Compressive characterization for building Direction 2 for the specimen production
- Bending characterization for building Direction 2 for the specimen production
- Tensile characterization for building Direction 3 for the specimen production
- Compressive characterization for building Direction 3 for the specimen production
- Bending characterization for building Direction 3 for the specimen production

The tensile characterization for the first building direction has been performed by the same authors in [16]. Some preliminary information about the compressive characterization for the first building direction has already been proposed in [17]. The complete and exhaustive compressive test for ABS specimens built in Direction 1 is the topic of this new work. Interesting studies and characterization tests of structural elements including ABS and produced via additive manufacturing have been proposed in [18–22].

A capability study focused on the compression analysis of specimens built with a Sharebot NG 3D Printer will be presented. Furthermore, a capability analysis on the dimensional properties, which measures the dimensional characteristics of the same specimens, will also be performed. The proposed statistical theory is based on the the the Six Sigma Process.

2. The Compression Test

The aim of this experimental campaign is to determine the compressive properties of ABS specimens produced with the FDM technique implemented in a home desktop Sharebot NG printer with a single extruder. The reference standard for the determination of the compressive properties of plastic materials (reinforced or not) is the ASTM D695 [23]. In the case of isotropic materials, at least five specimens must be tested. For this reason, nine specimens have been employed in the present test.

This standard proposes two types of specimen, depending on the properties to be determined. When compressive strength is desired, the specimens must have the form of a right cylinder or prism where the length is twice its principal width or diameter. However, when compressive elastic modulus and offset yield-stress data are investigated, the geometrical dimensions of the specimens are expressed in terms of the slenderness ratio. The slenderness ratio is defined as the ratio between the length of a column of uniform cross-section and its least radius of gyration (0.25-times the diameter for specimens of a uniform circular cross-section or 0.289-times the smaller cross-sectional dimension for specimens of a uniform rectangular cross-section). The proposed slenderness ratio should be in the range from 11:1 to 16:1. The second kind of specimen, with a rectangular section, has been chosen for the present study of the compressive properties. The reasons have been explained in Section 2.1.

2.1. Specimen Characteristics

The specimen proposed by the standard evolves along a principal direction. Figure 2 shows two of the three different possible building directions. The first one would be preferable, being coherent with the direction chosen in the tensile test performed by the authors in [16]. However, the choice of a circular cross-section specimen creates some differences in the printing process, and the fused material needs an appropriate support. For this reason, a support material basin could be planned as shown in Figure 3.

For the first printing direction, the support material does not completely adhere, and the specimens separate from the printing floor. For the second building direction, there are some difficulties, such as the small contact section between the specimen and the floor and the presence of high vibrations. These vibrations induce the separation of some specimens during the printing process. For all of these reasons, square cross-section specimens have been chosen here. The contact area of a square section is greater than that of a circle section with the same principal dimensions. Therefore, this type of specimen provides a greater adhesion. The dimensions were set to have a slenderness ratio, close to the lower values (from 11:1 to 16:1) proposed in the previous section, which avoids similar issues to the ones previously mentioned. The use of a square cross-section allows the use of the same printing direction employed for the tensile ABS characterization proposed in the work [16]; this direction and the specimen are indicated in Figure 4.

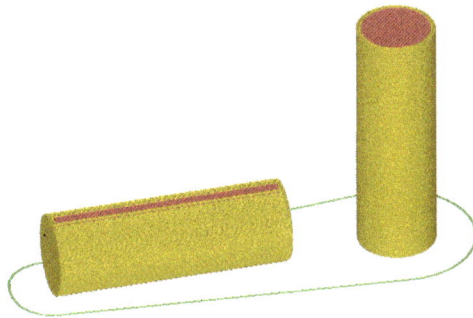

Figure 2. Possible building directions for ABS cylindrical specimens used in compression tests. First printing direction on the left and second printing direction on the right.

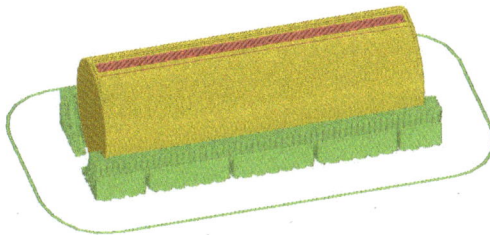

Figure 3. The support material basin for the construction of ABS cylindrical specimens along the first printing direction.

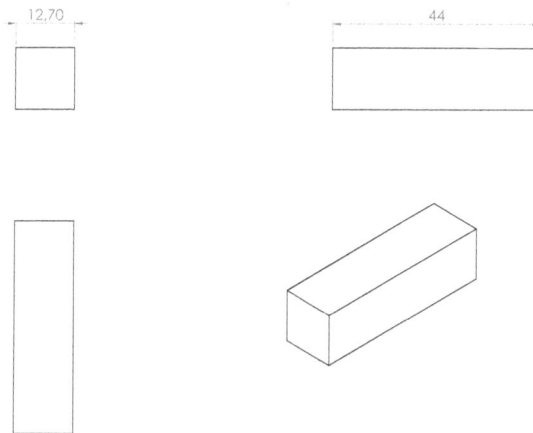

Figure 4. First printing direction and dimensions of the employed specimen with a square section.

2.2. Printing Parameters

The mechanical properties of 3D printed pieces are influenced by several printing parameters. These values must be chosen before generating the file containing the instructions to be followed by the printer. In the case of specimens for compression tests, Ahn et al. [24] examined only the building direction as a printing parameter, determining that a specimen with a building direction transverse to the compression direction would have shown a higher compressive strength. Since the influence of the other parameters was not investigated in the literature, it was chosen to set the printing parameters consistently with the tensile test already performed in [16]. Hereinafter, these parameters will be presented, together with their numerical values chosen by the authors. A specific explanation for each of these parameters can be found in [16]:

- Bead/Raster width: It is related to the nozzle gap as it represents the transverse dimension of the extruded bead. The nozzle size of the Sarebot NG is 0.35 mm [25]. Therefore, this fixed value was set.
- Perimeters: It represents the peripheral beads to be deposited. Two perimeter walls were used in the present work.
- Air gap: It is used to set the distance between two adjacent deposited beads in order to specify the internal infill density, which was set to 100%. The aim is to obtain solid specimens without overlapping beads.
- Bed temperature: The print plane was heated to 90 °C to prevent the deformation of specimens after a quick cooling.
- Build temperature: ABS is commonly extruded in a temperature range between 220 °C and 250 °C. Therefore, a nozzle temperature of 245 °C was set here.
- Raster orientation/angle: When the fill-pattern is rectilinear, it indicates the orientation angle of the filling beads. Crisscross specimens were here printed with a lamination sequence of 45°/−45°.
- Layer height: it measures the vertical dimension of each extruded bead. It was set to 0.2 mm.

2.3. Test Setup

Before starting the experimental campaign, it is compulsory to measure the width and thickness of each specimen to the nearest 0.01 mm, at several points along its length, recording the minimum value of the cross-sectional area. The length of each specimen must be also measured. Beyond the specific reasons for which such measurements are made, they can be useful in determining some "best practices" for the design phase to ensure the dimensional compatibility of the pieces to be printed, taking into account the errors introduced in the printing process. The measurements shown in Table 1 were made using a Burg Wachter PRECISE PS 7215 digital caliper, whose measuring range and accuracy are 150 mm and 0.01 mm, respectively [26]. Thickness and width values refer to the smallest section. The first specimen has been discarded because it has not given satisfactory results.

Table 1. Dimensional experimental data for the nine produced ABS specimens.

Specimen	2	3	4	5	6	7	8	9	10	*Nominal*	*Mean*	*StDev*
X dimension (mm)	12.70	12.73	12.74	12.72	12.83	12.73	12.71	12.76	12.79	12.70	12.75	0.04157
Y dimension (mm)	12.72	12.74	12.71	12.78	12.78	12.78	12.74	12.74	12.79	12.70	12.75	0.02958
Length (mm)	43.89	43.86	43.82	43.70	43.81	43.67	43.73	43.62	43.79	44.00	43.77	0.09071
Weight (g)	6.923	6.880	6.857	6.926	6.950	6.956	6.954	6.955	6.987	7.806	6.932	0.04090

Note: Nominal values (*Nominal*), mean values (*Mean*) and Standard Deviation (*StDev*).

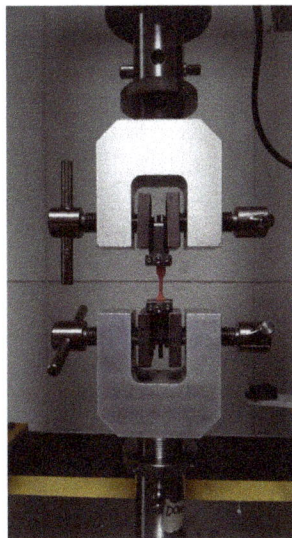

Figure 5. Artisan QTest10 [27] testing machine during an experimental tensile test.

An MTS QTEST 10 testing machine [27] (see the example in Figure 5 proposed for the tensile test) was used; each specimen was placed between an upper and a lower planar support taking care of the fact that the surfaces of the compression tools were parallel with the end surfaces of the specimen and aligning the center line of its long axis with the center line of the machine. As the common testing machines can be controlled in speed or in load, the standard requires that the test takes place in speed control, with a constant speed of the movable support of 1.3 mm/min. It is also suggested to increase the speed after the yield point, but only if the machine has a weighing system with rapid response and the material is ductile. However, it was preferred to maintain constant the load application speed for simplicity. In the specific case of this testing machine, the absolute position is that of the upper support.

3. Numerical Analysis

The raw stress-strain curves for the nine selected specimens are shown in Figure 6 where the compression behavior of each specimen is provided.

All of the proposed specimens show the common linear-elastic region, followed by a non-linear zone. Until about the proportional limit, the specimens show no macroscopically appreciable deformations. Subsequently, a first bulge of the central section appeared, followed by a progressive buckling of the specimen. Figure 7 shows the different modes of deformation that can occur in a compression test.

The various modes shown in Figure 7 can be summarized as follows [28,29]:

1. typical buckling mode: it happens when the ratio between the sample length and its width exceeds five;
2. shearing mode: it may happen when the ratio between the sample length and its width is about five;
3. double barreling mode: it may happen when the ratio between the sample length and its width exceeds two and friction is present at the contact surfaces;
4. barreling mode: it happens when the ratio between the sample length and its width is less than two;
5. homogenous compression mode: it happens when the ratio between the sample length and its width is between 2.0 and 1.5;
6. compression instability mode.

The tested specimens have a slenderness ratio of 12 (as will be calculated in Section 3.2). The ratio between the length and the width is 3.5. For this type of specimen, the standard [23] does not provide for the use of a support jig to avoid the buckling of the specimen. Even if the ratio between the length and the width is 3.5, all of the specimens, after an initial deformation, which can be assumed as Mode 4, moved to the typical buckling mode. After reaching a maximum stress value in buckling mode, all of the specimens started to break up into zones with fibers subjected to tensile and shearing stresses. Figure 8 details the deformation evolution during the compression test. Figure 9 shows a typical broken specimen where fibers subjected to tensile and shearing stresses are mentioned.

3.1. Post Processing According to ASTM 695

Each datum collected from the compression test has been treated according to ASTM 695 (technically equivalent to the ISO 604) [23] in absence of a specific normative for FDM 3D printed objects. ASTM 695 norms collect the standard methods of test for the compressive properties of rigid plastics, un-reinforced or reinforced, including also composites, when loaded in compression at low uniform rates. Each of the stress-strain curves shows in the linear-elastic region a horizontal tangent point, spacing two sections with slightly different slopes. This toe region is an artifact caused by the take up of slack and alignment or seating of the specimen. As it does not represent a property of the material, it was compensated drawing a continuation of the linear region of the curve until the zero-stress is reached and considering the intersection of this straight line with the strain axis as the correct zero strain point. The compressive properties that can be determined by this experimental campaign are:

- Compressive modulus of elasticity: The coefficients of the linear regressions based on point-by-point increasing ranges of values were averaged. This procedure had as the starting point the one next to the graph change of slope and as the ending point the one at which the new regression coefficient differed by more than 5% from the averaged one.
- Compressive proportional limit: From the previously found modulus of elasticity, it was possible to identify the stress value, which differed by more than 5% from the expected value; this was conventionally identified as the proportional limit
- Maximum compressive stress: It was calculated by dividing the maximum load by the minimum cross-sectional area value. However, as all of the specimens suffered buckling, it is not advisable to take account of these values as compressive strength, and it will be necessary to repeat the test with more stubby specimens in accordance with the standard. The results are reported in any case to be thorough.

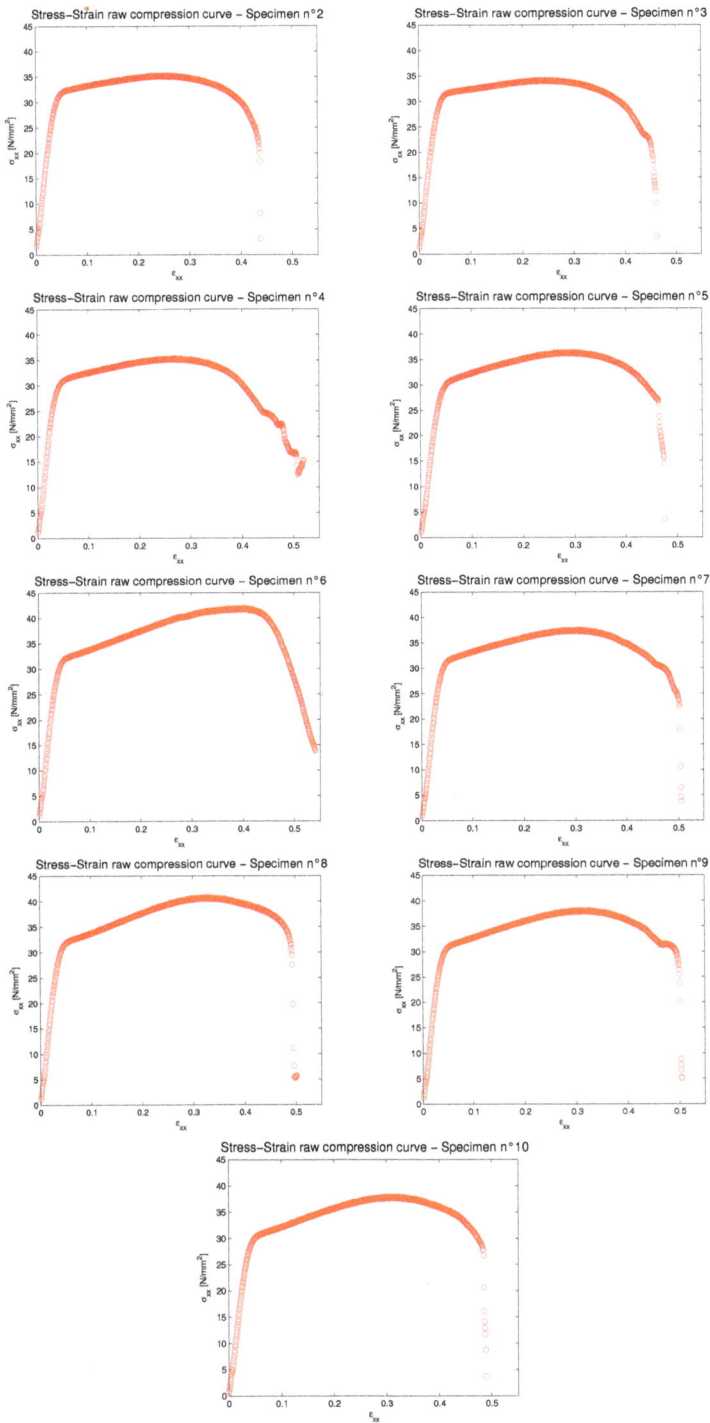

Figure 6. Raw stress-strain curves for the nine produced specimens.

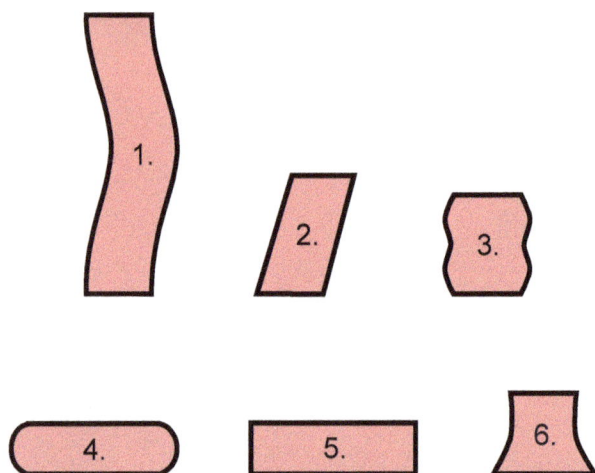

Figure 7. The possible deformation modes in a compression test.

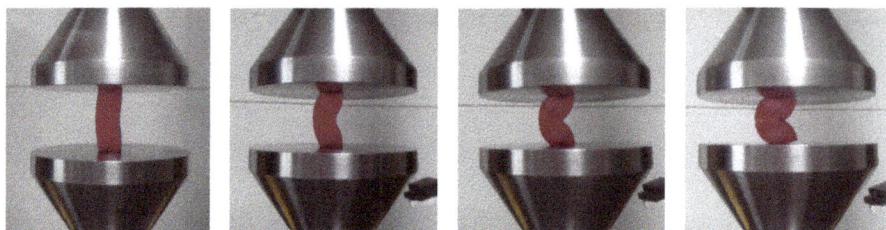

Figure 8. The behavior of the fourth specimen during the compression test.

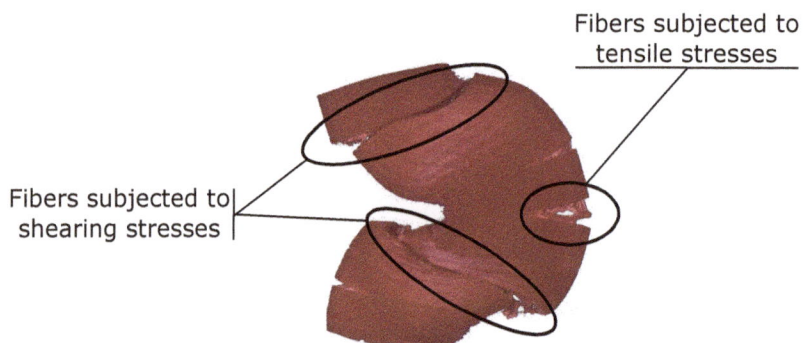

Figure 9. Condition of the fourth specimen after the compression test.

The application of corrections to the the linear portion of the stress-strain curve was not necessary in this work. The plots for each of the nine specimens are presented in Figures 10–18. Table 2 gives the collected results already shown in Figures 10–18.

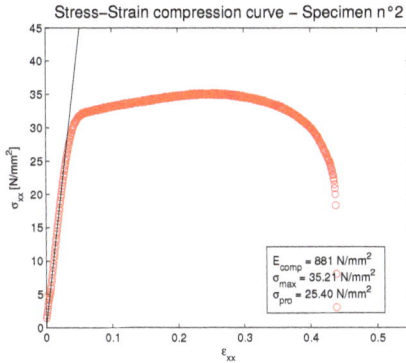

Figure 10. Actual stress-strain curve for the produced Specimen 2.

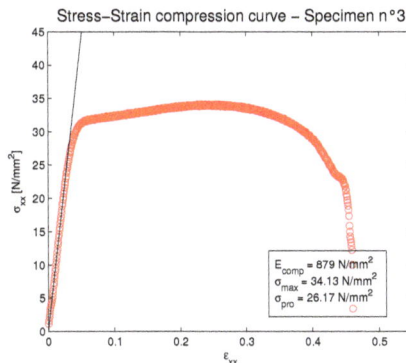

Figure 11. Actual stress-strain curve for the produced Specimen 3.

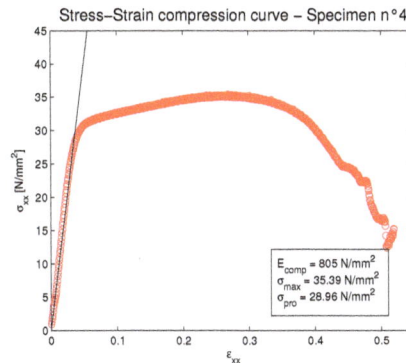

Figure 12. Actual stress-strain curve for the produced Specimen 4.

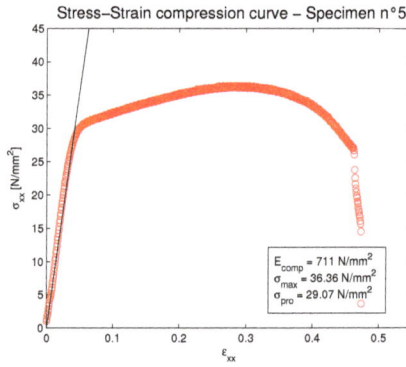

Figure 13. Actual stress-strain curve for the produced Specimen 5.

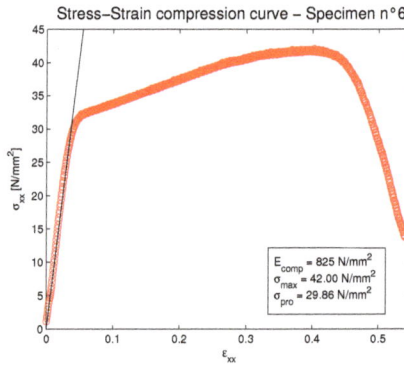

Figure 14. Actual stress-strain curve for the produced Specimen 6.

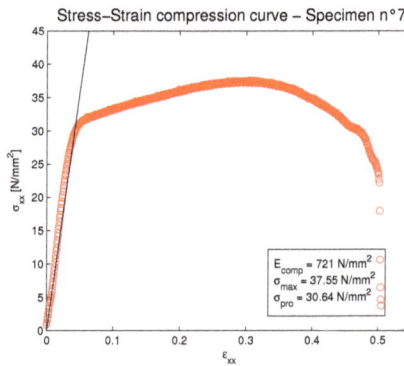

Figure 15. Actual stress-strain curve for the produced Specimen 7.

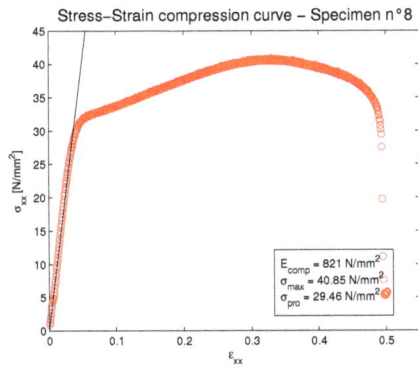

Figure 16. Actual stress-strain curve for the produced Specimen 8.

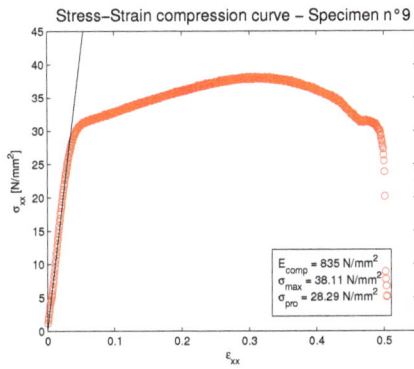

Figure 17. Actual stress-strain curve for the produced Specimen 9.

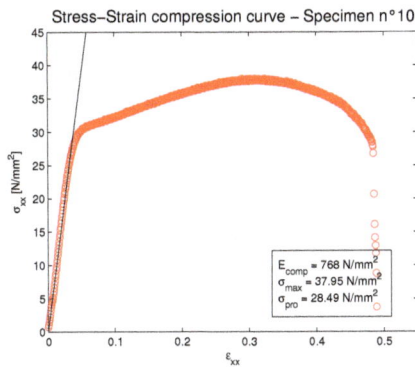

Figure 18. Actual stress-strain curve for the produced Specimen 10.

Table 2. Mechanical experimental data for the compression test of the 9 produced specimens.

Specimen	2	3	4	5	6	7	8	9	10	*Mean*	*StDev*
E (MPa)	881	879	805	711	825	721	821	835	768	805.1	61.31
σ_{max} (MPa)	35.21	34.13	35.39	36.36	42.00	37.55	40.85	38.11	37.95	37.51	2.608
σ_{pro} (MPa)	25.40	26.17	28.96	29.07	29.86	30.64	29.46	28.29	28.49	28.49	1.697

Note: Mean values (*Mean*) and Standard Deviation (*StDev*).

3.2. Compression Critical Load

The critical load is defined as the maximum load that a column can bear while staying straight. As before buckling, the specimens show no macroscopic deformations, it was interesting to investigate the theoretical value of the compression critical load to verify that it is higher than the proportional limit. As the slenderness ratio of a column increases, the critical load first follows the parabolic Johnson formula, then the hyperbolic Euler one [30]; the transition point can be expressed in the form of the slenderness ratio, imposing the tangency between the two curves [30]. The slenderness ratio can be calculated as:

$$A = \left(\frac{l_e}{\rho_g}\right)_{trans} = \sqrt{\frac{2\pi^2 E}{\sigma_y}} \qquad (1)$$

where l_e is the effective length, ρ_g the radius of gyration, E the tensile elastic modulus and σ_y the tensile maximum stress (it is obtained from past authors' work [16] where it is clear how the maximum tensile stress is very close to the tensile yield stress; this last stress has not been calculated in [16]).

The specimens will be analyzed as approximating pinned-pinned columns. Therefore, the effective length coincides with the real one, making A coincide with the slenderness ratio defined at the beginning. Taking into account the ABS tensile elastic modulus and the maximum stress found in [16], this formula leads to a slenderness ratio of 36.98. As the specimens' slenderness ratio (calculated as $\frac{l_e}{\rho_g}$) is 12, they can be interpreted as intermediate length columns, so the Johnson parabolic transition formula should be used:

$$\sigma_{cr} = \sigma_y - \frac{(\sigma_y A)^2}{4\pi^2 E}. \qquad (2)$$

This calculation, using a maximum tensile stress of 35.47 N/mm^2 in place of the yield stress (see the past authors' work [16]), gives a critical stress of 33.60 N/mm^2, which is slightly higher than the proportional limit and lower than the compressive strength. This feature means that all specimens collapsed for buckling reasons. However, results for elastic modulus E and proportional limit σ_{pro} are valid. On the contrary, new specimens with a different A value must be produced to obtain a higher value for the critical stress in order to correctly identify the maximum compressive stress. In Equations (1) and (2), the employed Young modulus $E = 2458$ N/mm^2 is the mean value obtained in the tensile test performed in [16]. This choice has been made because it is more conservative for the calculation of σ_{cr}. The tensile test in [16] proposed a proportional limit equal to 27.79 N/mm^2 (mean value).

3.3. Statistical Analysis

As already done for the tensile properties determined in [16], the mechanical and dimensional experimental values are here evaluated for compressive tests setting up a capability analysis. Since the mechanical properties and the geometrical values (in the sense of the printing deviations from the nominal values) are determined, the capability analysis is implemented to determine the upper and lower limits of these quantities, which can limit in a statistically stable way the experimental values.

As a precondition, it is necessary to verify if the investigated quantities could be approximated with a normal distribution. The Anderson–Darling hypothesis test [31,32] measures how well the data follow a particular distribution considering the values of two indices, which are the AD (Anderson–Darling statistic) and the p-value. For a specific set of data and a number of different distributions, the better fit is obtained for the smaller value of AD. The probability index should be as high as possible. A reference α value of 0.05 or 0.1 usually allows excluding or considering a certain distribution.

A goodness of fit test is set up for all of the experimental quantities shown in Tables 1 and 2. These results are given in Tables 3 and 4. It is necessary to verify if the Anderson–Darling statistic of a certain distribution was substantially smaller than the ones of the others and that, simultaneously, the correspondent p-value is higher than the reference value. The normal distribution does not always seem to be the best fit. However, both the indices allow its use. This analysis is discussed in detail in the next section.

Table 3. Individual distribution identification for the compression modulus of elasticity, the compressive stress at rupture σ_{max} and the compressive proportional limit σ_{pro}.

Goodness of Fit Test	Compression Modulus		σ_{max}		σ_{pro}	
	AD	p-Value	AD	p-Value	AD	p-Value
Normal	0.310	0.487	0.280	0.552	0.432	0.234
Box–Cox transformation	0.261	0.614	0.188	0.865	0.424	0.245
Lognormal	0.339	0.410	0.251	0.649	0.480	0.172
3-Parameter lognormal	0.348	–	0.207	–	0.460	–
2-Parameter Exponential	0.839	0.068	0.467	>0.250	1.256	0.015
Weibull	0.297	>0.250	0.429	>0.250	0.261	>0.250
N3-parameter Weibull	0.297	0.487	0.250	>0.500	0.239	>0.500
Smallest extreme value	0.297	>0.250	0.474	0.221	0.236	>0.250
Largest extreme value	0.477	0.218	0.213	>0.250	0.683	0.062
Gamma	0.367	>0.250	0.288	>0.250	0.494	0.228
Logistic	0.319	>0.250	0.279	>0.250	0.393	>0.250
Loglogistic	0.344	>0.250	0.258	>0.250	0.434	0.228
3-Parameter Loglogistic	0.319	–	0.209	–	0.393	–

Table 4. Individual distribution identification for the width, the thickness, the length and the weight of the produced specimens.

Goodness of Fit Test	Width: X Dimension		Thickness: Y Dimension		Length		Weight	
	AD	p-Value	AD	p-Value	AD	p-Value	AD	p-Value
Normal	0.439	0.223	0.574	0.097	0.191	0.855	0.474	0.179
Box–Cox transformation	0.424	0.245	0.575	0.096	0.090	0.858	0.461	0.195
Lognormal	0.437	0.227	0.573	0.097	0.191	0.853	0.478	0.175
3-Parameter lognormal	0.149	–	0.646	–	0.225	–	0.507	–
Exponential	–	–	–	–	4.112	<0.003	4.086	<0.003
2-Parameter exponential	0.285	>0.250	0.798	0.079	0.821	0.073	1.279	0.013
Weibull	0.685	0.061	0.712	0.049	0.196	>0.250	0.357	>0.250
3-Parameter Weibull	0.235	>0.500	0.627	0.092	0.215	>0.500	0.355	0.346
Smallest extreme value	0.688	0.060	0.712	0.049	0.196	>0.250	0.355	>0.250
Largest extreme value	0.237	>0.250	0.564	0.133	0.291	>0.250	0.674	0.066
Gamma	0.460	>0.250	0.645	0.094	0.225	>0.250	0.509	0.214
3-Parameter gamma	–	–	–	–	1.071	–	2.801	–
Logistic	0.381	>0.250	0.619	0.062	0.227	>0.250	0.461	0.197
Loglogistic	0.379	>0.250	0.619	0.062	0.228	>0.250	0.464	0.194
3-Parameter loglogistic	0.151	–	0.619	–	0.227	–	0.461	–

4. Results

This section proposes the results for the capability analysis. Such an analysis has been performed for the three investigated mechanical properties and for the dimensional characteristics, including also the weight of the produced specimens.

4.1. Capability Analysis for Mechanical Properties

In this section, the mechanical properties are investigated. The quantities which are taken into account are those presented in Section 3.1: the compression elastic modulus (*E*), the compression proportional limit (σ_{pro}) and the compression strength (σ_{max}). The sample was composed by 10 specimens, but the first one was used as a sacrificial specimen to understand the machine operation. Therefore, the experimental values of the nine employed specimens are reported in Table 2. The analysis is carried out by means of the control chart, the probability plot and the capability analysis, which are proposed for each quantity.

The first line of Table 2 shows the experimental collected values for the Young modulus E. The probability plot in Figure 19 shows that the average value of the overall sample is equal to 805.1 N/mm^2, and the standard deviation is 61.31. The Anderson–Darling statistic value is equal to 0.310, which is not the smallest one among the considered distributions. However, being the p-value equal to 0.487, it is considerably higher than the threshold one (see Table 3). Therefore, the normal distribution can be considered a good fit for this set of data. Indeed, the data seem to follow approximately a straight line. The lower limit is 553.11 N/mm^2, and the upper limit is 1057.11 N/mm^2. To identify the upper and lower limits delimiting in a statistically stable way the percentage corresponding to the Sigma Level 4, a P_{pk} (Process Performance Adjusted for Process Shift index) equal to 1.33 was imposed; this would lead to a more conservative approach as it takes into account the long term variability, generally higher than the short one. However, as may happen while working with a small data sample size, P_{pk} appeared to be lower than the corresponding C_{pk} (Process Capability Adjusted for Process Shift index). Therefore, in this case, a C_{pk} equal to 1.33 was imposed. This feature results in a lower limit equal to 553.11 N/mm^2; at least 99.38% of the specimens will have a higher compression modulus of elasticity. The I-MR (Individuals (I) chart and Moving Range (MR) chart) chart presented in Figure 19 shows that the process is globally stable and performs inside of the limits of acceptance of the Sigma Level 4.

The experimental collected values for the compressive stress at rupture σ_{max} are given in the second line of Table 2. The overall sample mean is 37.51 N/mm^2, while the standard deviation is 2.608. From the probability plot in Figure 20, it can be deduced that the normal distribution can approximate this set of data well, as the Anderson–Darling statistic is 0.280 and the p-value is 0.552 (see Table 3). The capability report shows that the boundary limits for a P_{pk} index equal to 1.33 are 27.1056 MPa and 47.9096 MPa. The graph is quite symmetrical, although the left part is more populated. This capability analysis has been proposed anyway even if the critical load is smaller than the compression stress at rupture σ_{max}. This study has been reported only to be thorough.

The compressive proportional limit σ_{pro} is the last investigated mechanical characteristic. The experimentally-collected values are shown in the third line of Table 2. As can be seen in the probability plot of Figure 21, the data seem to follow a straight line, except for the the two smaller values. Indeed the Anderson–Darling statistic and the p-value suggest that the normal distribution approximates the difficulties of this dataset (see Table 3). The process capability report in Figure 21 underlines an important variability of this quantity, characterized by an average value of 28.49 MPa and a standard deviation of 1.697. The boundary limits were identified in 21.7012 MPa and 35.27 MPa imposing a Ppk index (Process Performance Adjusted for Process Shift index) equal to 1.33. The last graph of Figure 21 shows that, except for the first two specimens, the process is stable and inside of the boundaries.

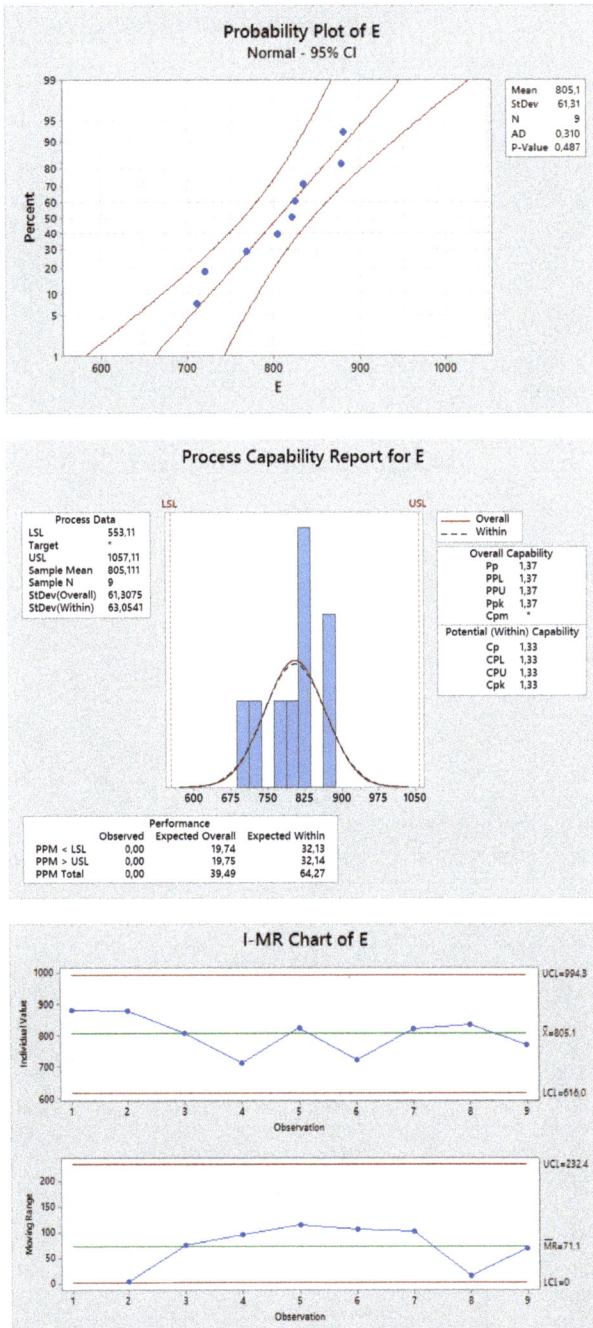

Figure 19. Probability plot, process capability report and I-MR chart (Individuals (I) chart and Moving Range (MR) chart) for the Young modulus *E* of the nine produced specimens.

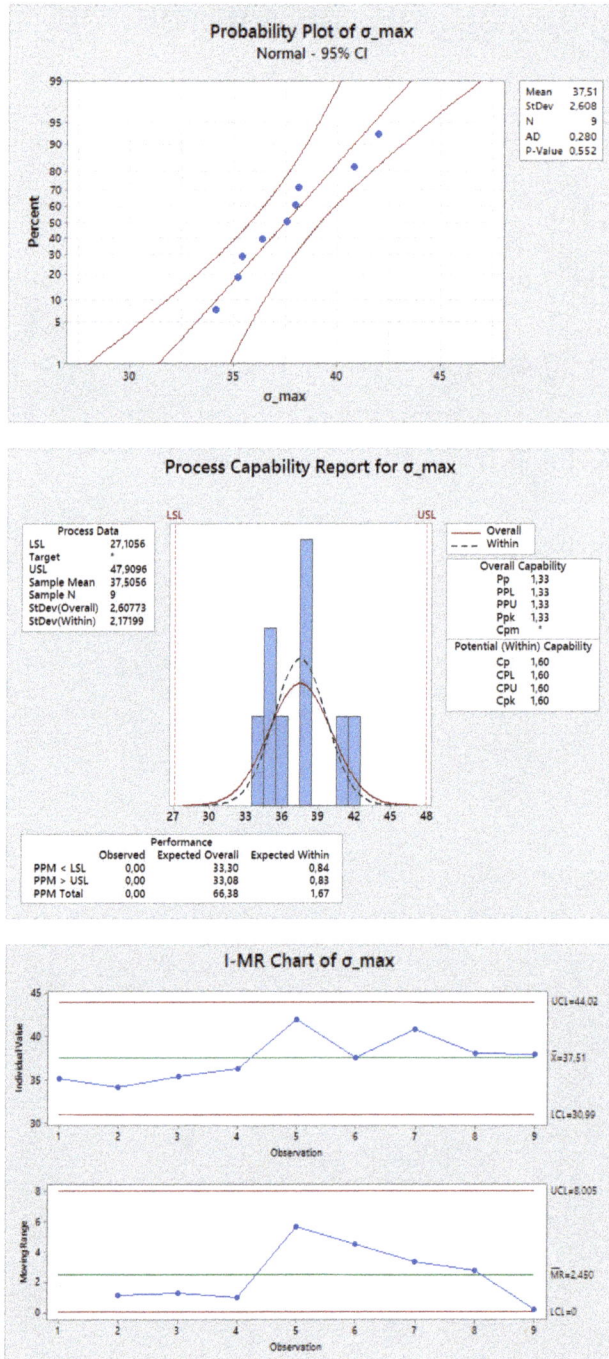

Figure 20. Probability plot, process capability report and I-MR chart (Individuals (I) chart and Moving Range (MR) chart) for the maximum stress at rupture (σ_{max}) of the nine produced specimens.

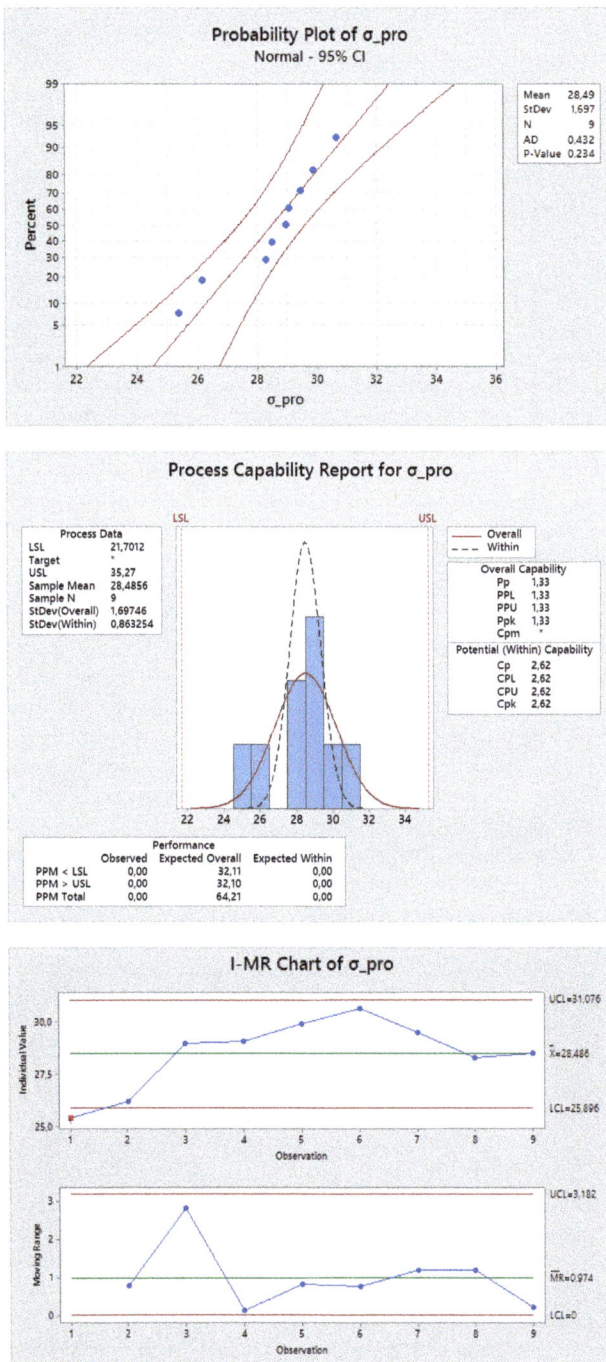

Figure 21. Probability plot, process capability report and I-MR chart (Individuals (I) chart and Moving Range (MR) chart) for the stress at proportional limit (σ_{pro}) of the nine produced specimens.

4.2. Capability Analysis for Dimensional Characteristics

In this section, the dimensional characteristics are investigated. Four quantities are taken into account: the X dimension and the Y dimension of the smallest section, the overall length and the weight (which shows the variability of the amount of extruded material). The sample was composed by nine specimens; all of the experimental values are given in Table 1. The analysis is carried out by means of the control chart, the probability plot and the capability analysis, which are presented for each of the four considered quantities.

The nominal value of the X dimension is 12.70 mm. The average real value is 12.75 mm, with a standard deviation of 0.04157. From the probability plot of Figure 22, it can be seen that the measurements seem to be shifted towards values higher than the nominal value. As the Anderson–Darling statistic is not so low and the *p*-value is 0.223, the normal distribution shows some difficulties in approximating this data distribution (see Table 4). However, the average percentage error on the X axis is just 0.4%, and this information can be introduced as a re-scaling factor. The lower and upper limits, delimiting the specimens' percentage corresponding to the Sigma Level 4, were identified, respectively, imposing a P_{pk} equal to 1.33, in 12.58 mm and 12.9112 mm. This approach takes into account the long-term variability, and it is, therefore, conservative, as it leads to a C_{pk} equal to 1.35.

The Y dimension parameter allows evaluating the printer's behavior along the Y axis. As for the X dimension, the nominal value is 12.70 mm. The average value is also 12.75 mm, as can be seen in the probability plot of Figure 23. This figure shows that also in this direction the printer manifests an average percentage error of 0.4%. However, the collected values do not follow a normal distribution, as can be seen from the extremely low *p*-value (see Table 4). Furthermore, the largest number of measures tends to be focused on values of 12.74 mm and 12.78 mm. It is therefore advisable to repeat the study on this axis, to better understand this behavior, trying to identify the possible existence of external effects. A P_{pk} value (Process Performance Adjusted for Process Shift index) of 1.33 was imposed, leading to a lower specification limit of 12.635 mm and an upper specification limit of 12.8716 mm. The capability histogram in Figure 23 shows an equally-spaced distribution. The in-plane performance of the printer seems to have the same percentage deviations from the nominal ones. However, further analysis should be carried out to identify the sources of the random behavior manifested in the Y dimension before taking into account a scale factor when the printing instructions are sent to the printer. The printer's behavior in the printing peripheral areas should also be deepened, as the specimens, having small dimensions, were always printed in the central area.

The length of the specimens allows evaluating the out-of-plane behavior of the printer. The design value is 44.00 mm. The probability plot of Figure 24 shows that the real dimension is always smaller than the nominal one. The average real value is 43.77 mm with a standard deviation of 0.09071. In this case, the process seems to be stable and controlled, as the Anderson–Darling statistic is significantly low, and the *p*-value is 0.855 (see Table 4). The capability histogram reveals that the process is well centered on the average value; indeed a lower specification limit of 43.4036 mm and an upper one of 44.1276 mm lead to similar P_{pk} and C_{pk} indices, respectively 1.33 and 1.40. The average error introduced in the printing process can be evaluated in −0.5%. This information is stable, and it can be used as a re-scaling factor with confidence. However, also in this case, the printer's behavior in printing peripheral areas should be studied.

The capability analysis on specimens' weight is made to determine how the amount of extruded material varies during the process. The specimen's volume was evaluated through its nominal dimensions; it should be equal to 7096.76 mm^3. The density of ABS filament declared by the vendor is 0.0011 g/mm^3. Therefore, a weight of 7.806 g is expected. Values shown in the last line of Table 1 indicate that the printer underestimated the amount of material to be extruded; indeed an average value of 6.932 g was found. This is consistent with what was found in the previous authors' work about the tensile characterization of ABS [16]. The behavior reported in Figure 25 is quite random, as the values do not seem to follow the straight line. Therefore, the Anderson–Darling statistic is low, and the *p*-value is just over the reference value (see Table 4). However, the lower limit is 6.769 g, and the upper limit is 7.095 g.

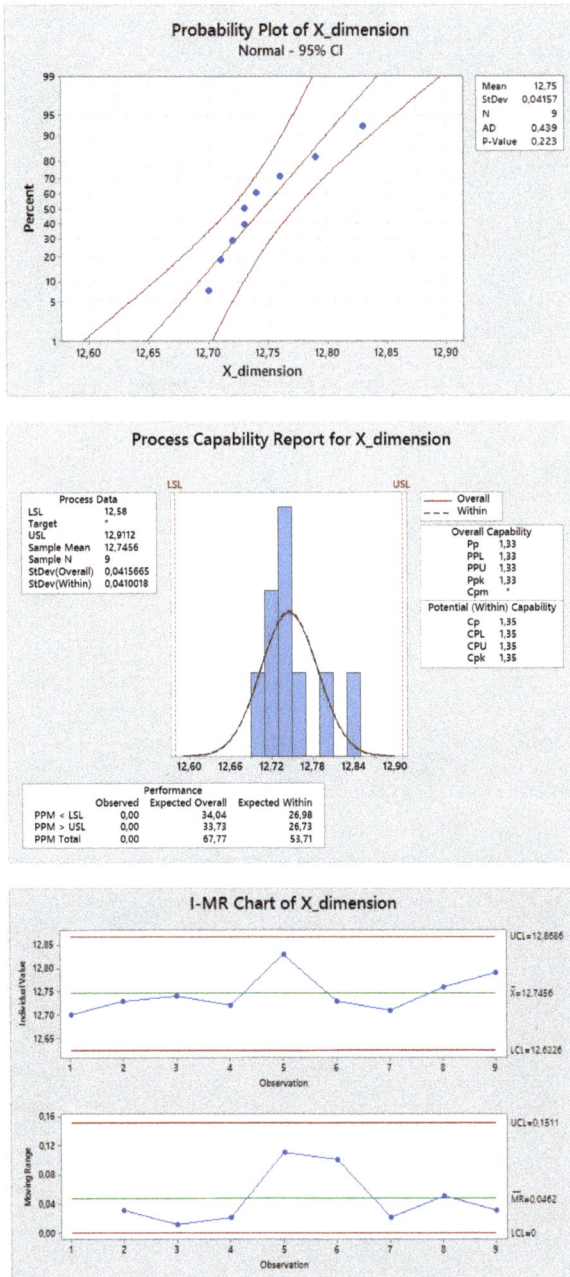

Figure 22. Probability plot, process capability report and I-MR chart (Individuals (I) chart and Moving Range (MR) chart) for the X dimension of the nine produced specimens.

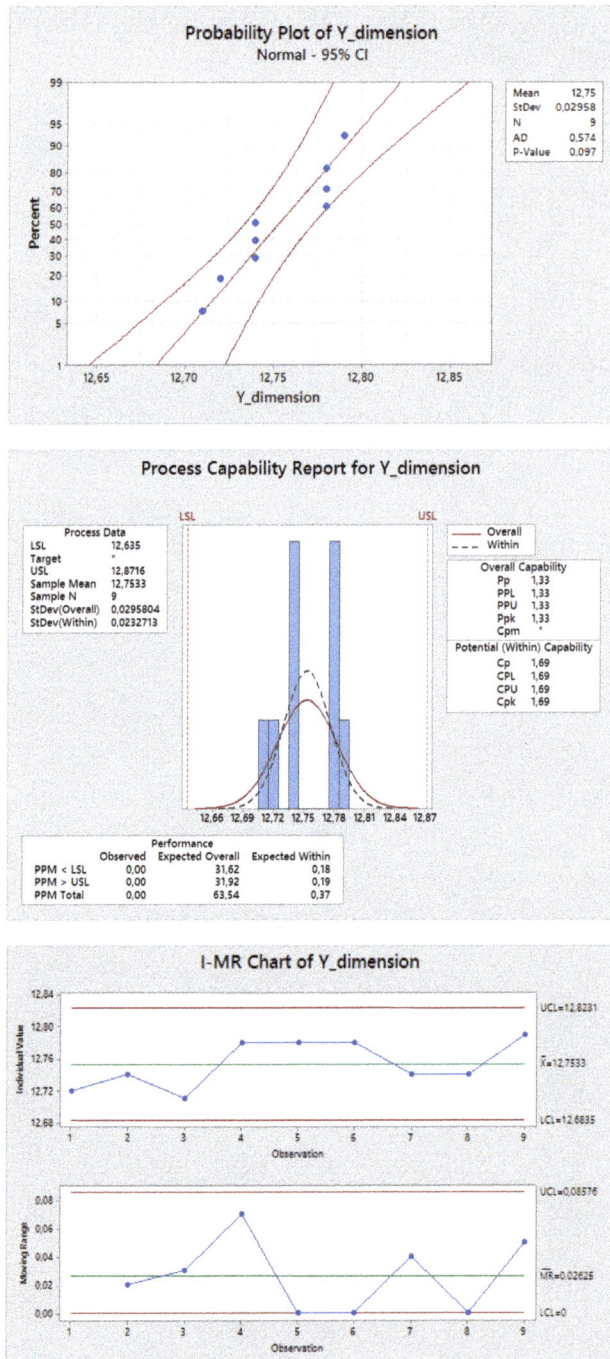

Figure 23. Probability plot, process capability report and I-MR chart (Individuals (I) chart and Moving Range (MR) chart) for the Y dimension of the nine produced specimens.

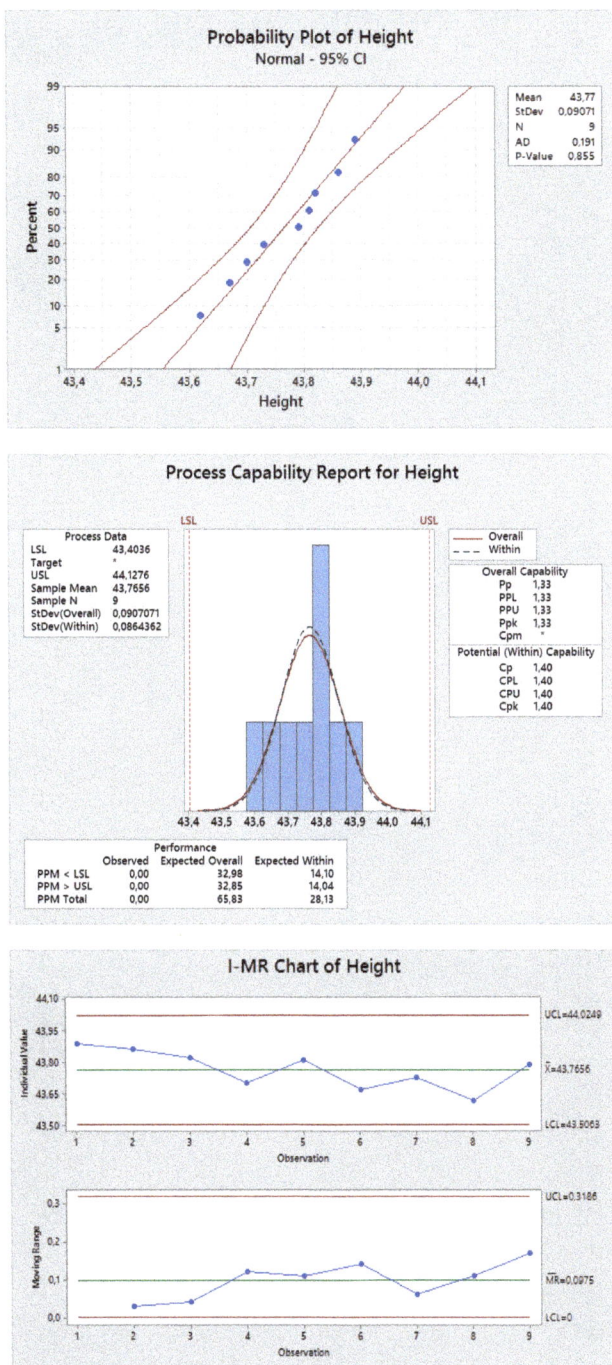

Figure 24. Probability plot, process capability report and I-MR chart (Individuals (I) chart and Moving Range (MR) chart) for the length of the nine produced specimens.

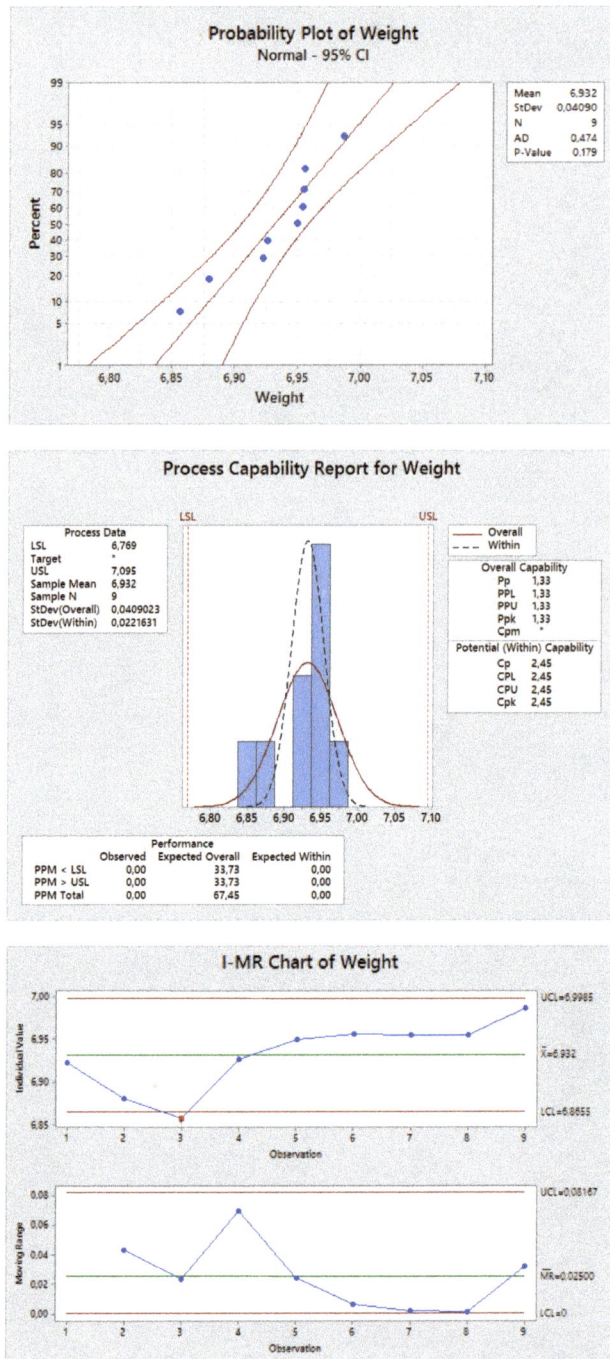

Figure 25. Probability plot, process capability report and I-MR chart (Individuals (I) chart and Moving Range (MR) chart) for the weight of the nine produced specimens.

5. Conclusions and Further Developments

The tensile properties of ABS specimens, printed with a desktop 3D printer, were studied in [16]. In order to fully characterize the mechanical properties of 3D-printed ABS, further analyses are necessary. One of these further analyses, here proposed, is the study of the compressive properties. A capability analysis, to evaluate if the 3D printing process is adequately stable for the self-production of flying components, has also been performed. A Sigma Level of 4 is chosen, so the lower limits for compressive elastic modulus, proportional limit and maximum stress were identified in 553.11 MPa, 21.7012 MPa and 27.1056 MPa, respectively. The found proportional limit and maximus stress are consistent with the compressive strength found in [24]. Being sufficiently conservative, these values can be used with confidence as input for a structural analysis. It is interesting to note that the critical load value is consistent with the experimental evidence. As stated before, the specimen shows some macroscopically appreciable deformation at a load level higher than the proportional limit. The ideal critical load resulted in being higher than the proportional limit. It is advisable to repeat the compression test with more stocky specimens (in order to avoid the buckling mode), to correctly evaluate the compressive maximum stress here wrongly calculated as 27.1056 MPa. This experimental campaign can be completed with the bending test and with the repetition of tensile, compressive and bending tests for the other two printing directions.

Acknowledgments: The authors thank Raffaella Sesana for the assistance with the testing of the specimens and for advice that greatly improved the manuscript. We would also like to give our gratitude to all of the members of the PoliDrone team for their support and constant work.

Author Contributions: All authors contributed equally to this work.

Conflicts of Interest: The authors declare no conflict of interest.

1. Wright, D. Drones: Regulatory challenges to an incipient industry. *Comput. Law Secur. Rev.* **2014**, *30*, 226–229.
2. Clarke, R.; Moses, L.B. The regulation of civilian drones' impacts on public safety. *Comput. Law Secur. Rev.* **2014**, *30*, 263–285.
3. Clarke, R. The regulation of civilian drones' impacts on behavioural privacy. *Comput. Law Secur. Rev.* **2014**, *30*, 286–305.
4. ENAC Regulation, Remotely Piloted Aerial Vehicles. Available online: https://www.enac.gov.it/repository/ContentManagement/information/N122671512/Reg_APR_Ed202_2.pdf (accessed on 26 July 2015).
5. Montgomery, A. The Drone Economy Moves Beyond Science Fiction. Available online: https://www.forbes.com/sites/mikemontgomery/2015/11/05/the-drone-economy-moves-beyond-science-fiction/#3828b9022aa3 (accessed on 4 May 2017).
6. Brischetto, S.; Ciano, A.; Raviola, A. A Multipurpose Modular Drone with Adjustable Arms. Patent Temporary Number 102015000069620, 5 November 2015.
7. Ferro, C.G.; Grassi, R.; Seclì, C.; Maggiore, P. Additive Manufacturing Offers New Opportunities in UAV Research. *Procedia CIRP* **2015**, *41*, 1004–1010 .
8. Khajavi, S.H.; Partanen, J.; Holmstro, J. Additive manufacturing in the spare parts supply chain. *Comput. Ind.* **2014**, *65*, 50–63.
9. Brischetto, S.; Ciano, A.; Ferro, C.G. A multipurpose modular drone with adjustable arms produced via the FDM additive manufacturing process. *Curved Layer. Struct.* **2016**, *3*, 202–213.
10. Rossi, G.; MORETTI, S.; CASAGLI, N. An Improved Drone Structure. Patent WO2015036907A1, 19 March 2015.
11. Choe, J.P. The Rotor Arm Device of Multi-Rotor Type Drone. Patent KR101456035B1, 4 November 2014.
12. Guilhot-Gaudeffroy, M. Gyroptere a Securite Renforcee. Patent WO2004113166A1, 29 December 2014.
13. Desaulniers, J.-M. Amphibious Gyropendular Drone for Use in e.g., Defense Application, has Safety Device Arranged in Periphery of Propulsion Device for Assuring Floatability of Drone, and Upper Propulsion Device for Maintaining Drone in Air During Levitation. Patent FR2937306A1, 23 April 2010.
14. Krache, R.; Debbah, I. Some mechanical and thermal properties of PC/ABS blends. *Mater. Sci. Appl.* **2011**, *2*, 404–410.

15. Nunez, P.J.; Rivas, A.; Garcia-Plaza, E.; Beamud, E.; Sanz-Lobera, A. Dimensional and surface texture characterization in fused deposition modelling (FDM) with ABS plus. *Procedia Eng.* **2015**, *132*, 856–863.
16. Ferro, C.G.; Brischetto, S.; Torre, R.; Maggiore, P. Characterisation of ABS specimens produced via the 3D printing technology for drone structural components. *Curved Layer. Struct.* **2016**, *3*, 172–188.
17. Brischetto, S.; Ferro, C.G.; Torre, R.; Maggiore, P. Tensile and compression characterization of 3D printed ABS specimens for UAV applications. In Proceedings of the 3rd International Conference on Mechanical Properties of Materials (ICMPM 2016), Venice, Italy, 14–17 December 2016.
18. Bellini, A.; Guceri, S. Mechanical characterization of parts fabricated using fused deposition modeling. *Rapid Prototyp. J.* **2003**, *9*, 252–264.
19. Anitha, R.; Arunachalam, S.; Radhakrishnan, P. Critical parameters influencing the quality of prototypes in fused deposition modelling. *J. Mater. Process. Technol.* **2001**, *118*, 385–388.
20. Weng, Z.; Wang, J.; Senthil, T.; Wu, L. Mechanical and thermal properties of ABS/montmorillonite nanocomposites for fused deposition modeling 3D printing. *Mater. Des.* **2016**, *102*, 276–283.
21. Torrado, A.R.; Shemelya, C.M.; English, J.D.; Lin, Y.; Wicker, R.B.; Roberson, D.A. Characterizing the effect of additives to ABS on the mechanical property anisotropy of specimens fabricated by material extrusion 3D printing. *Addit. Manuf.* **2015**, *6*, 16–29.
22. Faes, M.; Ferraris, E.; Moens, D. Influence of inter-layer cooling time on the quasi-static properties of ABS components produced via fused deposition modelling. *Procedia CIRP* **2016**, *42*, 748–753.
23. ASTM D695-02. Standard Test Method for Compressive Properties of Rigid Plastics. ASTM International, West Conshohocken, PA, 2002. Available online: www.astm.org (accessed on 4 May 2017).
24. Ahn, S.-H.; Montero, M.; Odell, D.; Roundy, S.; Wright, P.K. Anisotropic material properties of fused deposition modeling ABS. *Rapid Prototyp. J.* **2002**, *8*, 248–257.
25. Sharebot srl. Sharebot Next Generation User Manual. 2015. Available online: https://www.sharebot.it/downloads/NG/Manuale_NG.pdf (accessed on 4 May 2017).
26. Burg Wachter PRECISE PS 7215 Digital Caliper User Manual. Available online: https://www.burg.biz/.../BUW-0621-15-PRECISE-Web-EN.pdf (accessed on 30 September 2016).
27. QTest 10, Artisan Technology Group. Available online: https://www.artisantg.com/info/PDF_4D54535F51746573745F446174617368656574.pdf (accessed on 27 April 2016).
28. Kuhn, H.; Medlin, D. Mechanical Testing and Evaluation. 2000. Available online: http://app.knovel.com/hotlink/toc/id:kpASMHVMT2/asm-handbook-volume-08/asm-handbook-volume-08 (accessed on 4 May 2017).
29. Altenaiji, M.; Schleyer, G.K.; Zhao, Y.Y. Characterisation of Aluminium Matrix Syntactic Foams Under Static and Dynamic Loading. In *Composites and Their Properties*; Hu, N., Ed.; InTech: Rijeka, Croatia, 2012. ISBN 978-953-51-0711-8.
30. Schmid, S.R.; Hamrock, B.J.; Jacobson, B.O. *Fundamentals of Machine Elements*, 3rd ed.; CRC Press: Boca Raton, FL, USA, 2014.
31. Anderson, T.W.; Darling, D.A. Asymptotic theory of certain "Goodness of Fit" criteria based on stochastic processes. *Ann. Math. Stat.* **1952**, *23*, 193–212.
32. Anderson, T.W.; Darling, D.A. A test of goodness of fit. *J. Am. Stat. Assoc.* **1954**, *49*, 765–769.

technologies

Article

In-Built Customised Mechanical Failure of 316L Components Fabricated Using Selective Laser Melting

Andrei Ilie, Haider Ali and Kamran Mumtaz *

Centre for Advanced Additive Manufacturing, Department of Mechanical Engineering, University of Sheffield, Sheffield S1 3JD, UK; andreieilie92@gmail.com (A.I.); mep12ha@sheffield.ac.uk (H.A.)
* Correspondence: k.mumtaz@sheffield.ac.uk; Tel.: +44-114-222-7789

Academic Editors: Salvatore Brischetto, Paolo Maggiore and Carlo Giovanni Ferro
Received: 31 January 2017; Accepted: 21 February 2017; Published: 25 February 2017

Abstract: The layer-by-layer building methodology used within the powder bed process of Selective Laser Melting facilitates control over the degree of melting achieved at every layer. This control can be used to manipulate levels of porosity within each layer, effecting resultant mechanical properties. If specifically controlled, it has the potential to enable customisation of mechanical properties or design of in-built locations of mechanical fracture through strategic void placement across a component, enabling accurate location specific predictions of mechanical failure for fail-safe applications. This investigation examined the process parameter effects on porosity formation and mechanical properties of 316L samples whilst maintaining a constant laser energy density without manipulation of sample geometry. In order to understand the effects of customisation on mechanical properties, samples were manufactured with in-built porosity of up to 3% spanning across ~1.7% of a samples' cross-section using a specially developed set of "hybrid" processing parameters. Through strategic placement of porous sections within samples, exact fracture location could be predicted. When mechanically loaded, these customised samples exhibited only ~2% reduction in yield strength compared to samples processed using single set parameters. As expected, microscopic analysis revealed that mechanical performance was closely tied to porosity variations in samples, with little or no variation in microstructure observed through parameter variation. The results indicate that there is potential to use SLM for customising mechanical performance over the cross-section of a component.

Keywords: additive manufacturing; mechanical properties; customisation; selective laser melting

---

1. Introduction and Background

Research in the field of Selective Laser Melting (SLM) is broad, with many focused on parameter optimisation for achieving consistent high density parts and establishing a relationship between parameters and final part mechanical properties for specific materials [1–13]. Understanding the phenomenon of residual stresses building up in SLM parts due to high thermal gradients and strategies for minimising the residual stresses and their detrimental effects such as cracking and warping of parts are other areas of interest for researchers [14–21]. Analysis of post-processing operations such as hipping and heat treatment of additively manufactured parts has also been a point of interest for researchers as these can enable significant improvement in part density and mechanical properties [15,18,22,23]. Despite the increased interest and growth in SLM research, the complexity of this rapid solidification process has resulted in high value industries approaching this technology with caution [24]. In addition, the requirements for process repeatability, tolerances, and feedstock traceability have increased in recent years [25] and are understandably strict in sectors such as aerospace and biomedical orthopedics.

SLM of stainless steel alloys, particularly 316L, has been of major interest to researchers due to its high corrosion resistance, formability, strength, weldability, and biocompatibility. SS 316L is suitable for medical applications (implants and prosthesis), pharmaceuticals, architectural applications, fasteners, aerospace parts, marine and chemical applications, and heat exchangers [26,27]. SS 316L SLM parts in as-built conditions have exhibited fatigue properties similar to conventionally manufactured parts, mainly due to the high ductility even after SLM processing [7]. If the correct combination of process parameters are chosen, SLM steel samples usually exhibit improved tensile properties compared to conventionally manufactured steel samples [28].

Customised Mechanical Properties Using SLM

The layer by layer building methodology used within SLM offers the potential for mechanical properties to be controlled layer to layer. However, to date, limited research has been undertaken exploring this possibility. Theoretically this variation in mechanical properties can be generated through a number of methods. The first and most obvious is by geometry manipulation, designing with the assistance of computational software can allow designers to create structures that will have a specific mechanical response to external loading. This feature is not exclusive to additive manufacturing technologies and can be achieved through a variety of other manufacturing processes (however, complexity may be limited when using conventional processes). Secondly properties can be controlled through the introduction of variable materials (i.e., functionally graded materials), although this is challenging due to the potential for thermal expansion mismatch between materials leading to delamination of multi-material layers, it is also difficult to recycle and separate materials from a powder bed that consists of graded multi-materials. Thirdly, microstructure manipulation could be used to vary mechanical properties across the cross-section of parts, this may be achieved through adjustment of melting regimes employed at each layer [3,29]. However, due to the rapid solidification of material within SLM, it is often difficult to alter the microstructure of individual layers significantly from their standard fine dendritic form solely through parameter control (i.e., laser power, exposure time, etc.). Suitable SLM processing conditions (i.e., promoting high density components) and rapid solidification limit generation of sufficient variability within the thermal history of each layer to promote enough microstructural change to allow mechanical performance to vary significantly. Microstructures can be altered through multiple reheating strategies and powder bed pre-heating, but this tends to affect multiple layers across a component and precise layer to layer control will be challenging. Pre-heating the powder bed to higher temperatures can assist in delaying solidification and coarsen the microstructure, however this heating will affect the majority of layers across a component without permitting control over microstructural variation across specific layers. Finally, mechanical properties could be altered through the introduction of controllable features such as porosity across a component. This can be controlled through varying laser processing parameters layer to layer and inducing lack of fusion porosity. The mechanical properties can be artificially altered, making the areas across a component mechanically weaker in the tensile compared to a fully dense region, this can be designed across a part, promoting preferential deformation of parts or failure at strategic locations. Currently, when tensile testing SLM components the exact location of failure is random and highly unpredictable without the use of X-ray analysis. Customising the mechanical performance of an SLM sample may be particularly useful for applications requiring a fail-safe mechanism to minimise harm or respond appropriately to specific types of loading conditions, examples include pressure relief in pressure valves or blow-off panels used in enclosure. Designed premature failure may also be used to prevent more expensive or catastrophic failure from occurring further down the line.

2. Experimental Methodology

As an overview of the experimental methodology, processing parameters were adjusted so that variation in sample density could be attained. Having established what effect these adjusted parameters had, samples were created such that variation in mechanical properties (i.e., preferential point of failure) would be created across strategic locations within samples. Parts were tested for density, hardness, and tensile strength.

A Renishaw SLM 125 was used during investigations, this system uses a 200 W fibre laser to process metallic powdered feedstock within a purged argon atmosphere. Gas atomised SS AISI 316L (15–45 micron) was used as the feedstock material, its composition is shown in Table 1.

Table 1. Weight % Composition.

Element	Fe	C	Si	Mn	P	S	Cr	Ni	Mo	N	Cu	O
%Composition	Bal	0.012	0.6	1.25	0.012	0.005	17.8	12.9	2.35	0.04	0.03	0.0185

2.1. Sample Testing

10 × 10 mm cubes were fabricated for testing hardness using a Vickers hardness testing machine (BSEN ISO 6507-1:2005 [30]). The level of porosity in the specimen was estimated by area fraction analysis of representative micrographs/fields using a method based on ASTM E2109-01 (2007) and BS 7590:1992 [31,32]. Tensile test specimens were manufactured according to the ASTM E8 standard [33]. The specimens were tested on a Tinius Olsen H25K-S UTM Benchtop Materials Tester. A summary of key dimensions of the tensile test pieces are shown in Figure 1. The samples were built vertically and tested along this axis.

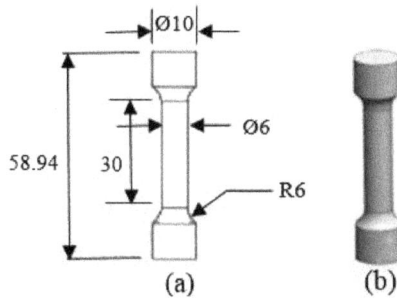

Figure 1. Specimen dimensions (**a**); and CAD file (**b**).

2.2. Processing Parameter Selection

For processing SS 316L a set of recommended laser process parameters were specified by SLM OEM Renishaw for creating components at full density, shown in Table 2.

Table 2. Manufacturer (Renishaw) recommended processing parameter values for SS 316L.

Parameter	Value
Layer Thickness—L (μm)	50 μm
Point Distance—x (μm)	50 μm
Hatch Spacing—h (μm)	90 μm
Spot Size (μm)	50 μm
Laser Power—P (W)	200 W
Exposure time—t (s)	70 μs

Various processing parameters affect the SLM process, these may be direct or indirect as specified by Yadroitsev et al. [2]. However, the principal parameters in SLM—those which have the most substantial effect—have been identified as being laser power, spot size diameter, exposure time, spot spacing (scan speed), hatch distance, and powder layer thickness [1]. Varying laser power and scan speed individually affects the laser energy density that is used to melt the powder bed during SLM, and controls levels of porosity within a component. This implies that there is an "optimum" energy density for fabricating fully dense parts free of irregularities as noted by Kurian et al. [10]. The laser energy density behaviour is described by Equation (1), and it was shown that laser power and exposure time are inversely proportional (other variables kept constant), in theory it is possible to vary both the laser power and exposure time in unison while maintaining the energy density delivered constant.

$$Energy\ density = Q = \frac{Pt}{xhl} \tag{1}$$

P: Laser power (W, J/s); t: Exposure time (s); x: Point distance (mm); l: Layer Thickness (mm); h: Hatch Spacing (mm); Q: Energy density (J/mm^3).

For these reasons, a variation of laser power with exposure time was performed, while maintaining an arbitrary "optimal" energy density. The energy density selected was 62 J/mm^3, consistent with the Renishaw SLM 125 advised processing parameters. This energy density was maintained in combination with varying exposure times and laser powers as shown in Table 3, continuous melt tracks were identified at powers as low as 150 W in other work [9]. Other SLM parameters were kept constant.

Table 3. Varied parameter range and processing set name/condition.

Specimen Set	A	B	C
Laser power (W)	200	175	150
Exposure time (μs)	70	80	93

Using the Renishaw SLM 125 and processing SS 316L powder, three repeat samples of each specimen type A, B, and C were produced as shown in Figure 2.

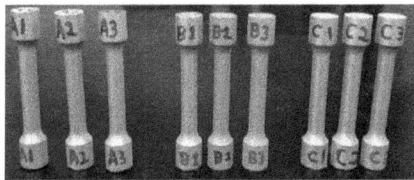

Figure 2. Uniform specimen sets.

The specimens were attached to the substrate plate via support structures in order to facilitate removal. Images of the support structures, including the surface texture of the specimen, with top and bottom (chiseled) sides are shown in Figure 3.

Figure 3. Support structures (**a**); and top (**b**); bottom (**c**) surface texture.

3. Results

Three sets of SS 316L cubed samples and tensile test bars were produced for each processing condition (A, B, C) using SLM parameters shown in Table 3. Based on these results, further customised samples were created and tested, their results are later detailed in Section 3.3.

3.1. Optical Microscopy

Microscopy was performed as described in Section 2.1, initially testing cubed samples for porosity and shown in Figure 4. Samples produced with parameter set A had an average density of 99.8%, parameter set B 99.1%, and parameter set C 96.8%. Even though energy density was maintained at 62 J/mm^3, as the laser power was reduced the porosity within samples increased due to a potential increase in lack of fusion, evident from the irregularly shaped pores. Subsequently, the etched samples were examined for a more detailed view of the microstructure shown in Figure 5, microstructures were typical of those produced using SLM with little or no variation in microstructure between samples when melting at different laser powers.

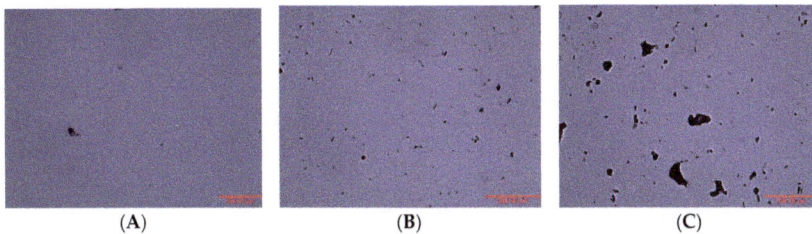

(A)　　　　　　　　**(B)**　　　　　　　　**(C)**

Figure 4. Porosity of cubes fabricated using parameters (**A**) (200 W); (**B**) (175 W); (**C**) (150 W).

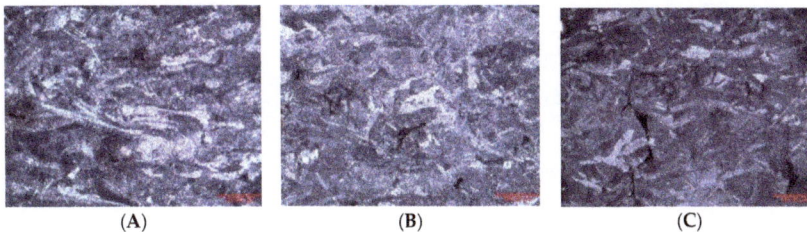

(A)　　　　　　　　**(B)**　　　　　　　　**(C)**

Figure 5. Microstructure of cubes fabricated using parameters (**A**) (200 W); (**B**) (175 W); (**C**) (150 W).

3.2. Tensile Testing

Tensile testing was performed on each specimen set (A, B, and C). The specimens fractured at varying points along the gauge length as can be seen in Figure 6. It can be seen that fracture location across the gauge length of samples are random and unpredictable.

Figure 6. Set B (175 W) specimen fracture locations.

Figure 7 shows stress-strain plots from tensile testing for one representative sample from each processing condition. It reveals that a 3% porosity variation between samples A–C processed at constant energy densities (but varying laser powers) was sufficient to generate variation in mechanical properties. As expected, trends generally indicate weaker parts (reduced yield strength, UTS, and fracture strength) were formed with use of lower power lasers due to higher levels of part porosity. The mean average for all samples (×3 repeats) is shown in Table 4 and Figure 8.

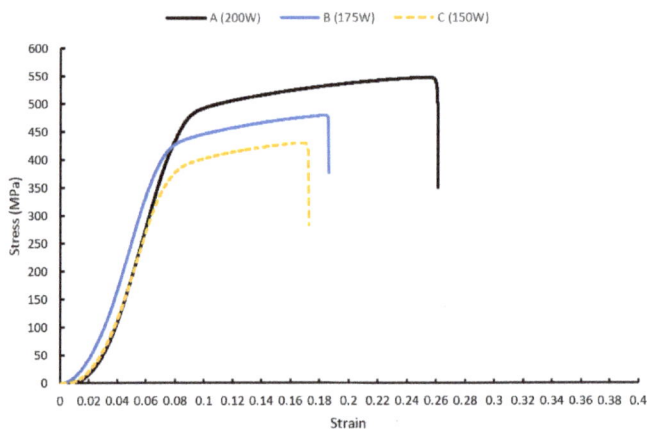

Figure 7. Uniform specimen stress-strain plot.

Table 4. Mean mechanical property values for uniform specimen sets.

Property \\ Specimen	A (200 W)	B (175 W)	C (150 W)
0.2% Yield point (MPa)	443 ± 1	385 ± 3	344 ± 5
UTS (MPa)	565 ± 17	483 ± 9	425 ± 22
Fracture stress (MPa)	558 ± 13	473 ± 10	405 ± 33

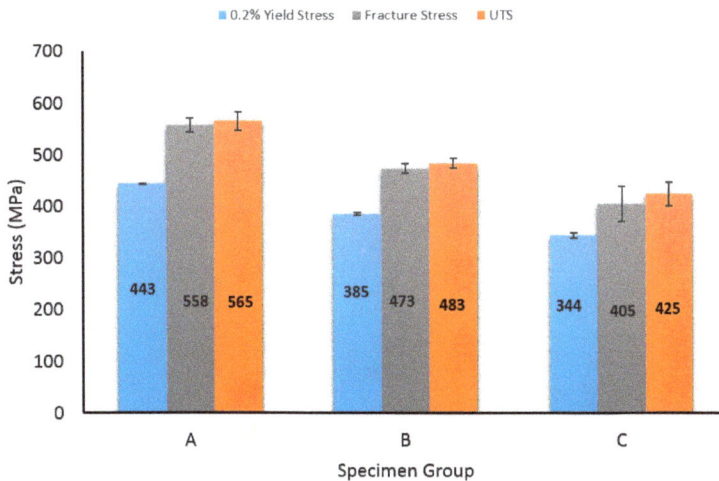

Figure 8. Comparison of mean mechanical properties between samples (**A**) (200 W); (**B**) (175 W); (**C**) (150 W).

3.3. Hardness Testing Results

Hardness tests were performed on cubic SLM samples. The mean average result for each sample is shown in Table 5. The higher apparent scan speeds (generated with shorter exposure times) used in sample A led to increased sample hardness due to faster solidification of the melt pool. Slower apparent scan speeds (as those used in sample C) lead to slower solidification rates and therefore a softer material. Higher hardness values generally indicate higher tensile strength and are therefore in agreement with the tensile results for each sample detailed in Section 3.1.

Table 5. Uniform specimen hardness testing results.

Sample	A	B	C
Average Vickers Hardness (HV)	193 ± 1	185 ± 3	159 ± 4

3.4. Customised Specimens

The results obtained for the uniform sample set show that there is significant variance in the mechanical properties as energy density remains constant and laser powers and exposure times are adjusted. The uniform sample set fractured in unpredictable locations within the gauge length (Figure 6). In order to customise mechanical properties and create predictable failure/break locations across specimens, two different sets of processing parameters were used to build a single tensile test specimen to initiate "structural change" within the sample. "Failure points" were introduced in the customised specimens at different locations along the gauge length of each sample as shown in Figure 9.

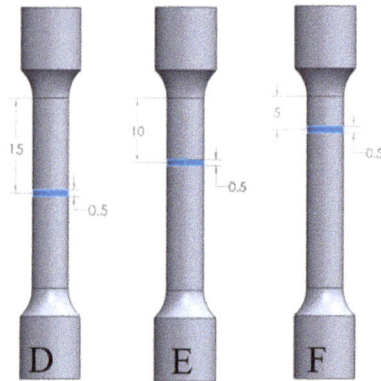

Figure 9. Designed failure locations for customised specimen set.

In order for the customised specimens to be produced on the SLM machine, each specimen was split into three parts; the bottom end, the top end, and the failure point. This distinction was made in order for the failure point to be assigned a different processing parameter set compared to the top/bottom ends. For the customised specimen sets D, E, and F (Figure 9), the failure points were placed at 15, 10, and 5 mm respectively from the top section of the gauge length. The failure point was designed to be 10 layers thick (i.e., 500 µm) to encourage an adequate or distinct change in mechanical performance. The processing parameters were 200 W laser power with 70 µs exposure time for the top/bottom ends of the sample (original parameter set A), and 150 W with 93 µs for the failure point (original parameter set C). These parameters were selected from the uniform sample sets A and C as they displayed the largest variation in mechanical properties as detailed in Section 3.2. The specimens produced are geometrically identical to the uniform tensile samples produced in Section 3.2, however, the failure point locations were labelled on the customised specimens for better visualisation, as shown in Figure 10.

Figure 10. Customised specimen set failure points.

Tensile tests were performed on the customised specimen sets D, E, and F. All samples fractured at the marked locations for designed in-built failure point, shown in Figure 11. This is evidence that predictable and controlled fracture/failure can be achieved through selection and implementation of multiple "hybrid" laser processing parameters within a single build.

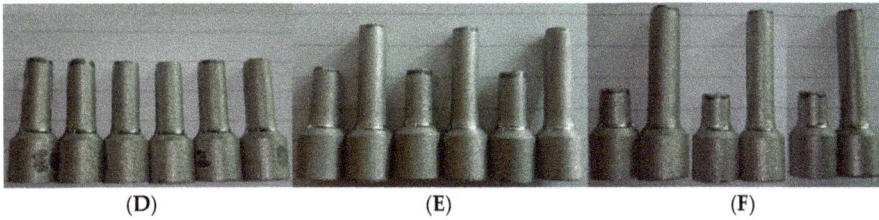

(D) (E) (F)

Figure 11. Customised specimen controlled fracture locations (**D–F**).

The mechanical properties of the sample sets D, E, and F are shown in Table 6 with the stress-strain plot for these samples shown in Figure 12. As expected, the mechanical properties of these customised parts were weaker than samples produced entirely with optimum parameter A (200 W) settings (Table 4), this is because approximately 1.7% of the SLM gauge length was produced using parameter set C (150 W) marginally increasing porosity within the component across 10 layers (500 µm). When comparing single parameters of set A samples, customised samples set D held approximately 4% lower yield strength and 10% lower UTS. Customised sample sets E and F showed a yield strength comparable to that of sample set A, while the UTS showed a reduction of approximately 5.5%.

Table 6. Mean mechanical property values for customised specimen sets.

Specimen Property	D	E	F
0.2% Yield point (MPa)	421 ± 12	443 ± 9	442 ± 12
UTS (MPa)	512 ± 7	536 ± 3	533 ± 9
Fracture stress (MPa)	509 ± 7	529 ± 3	526 ± 9

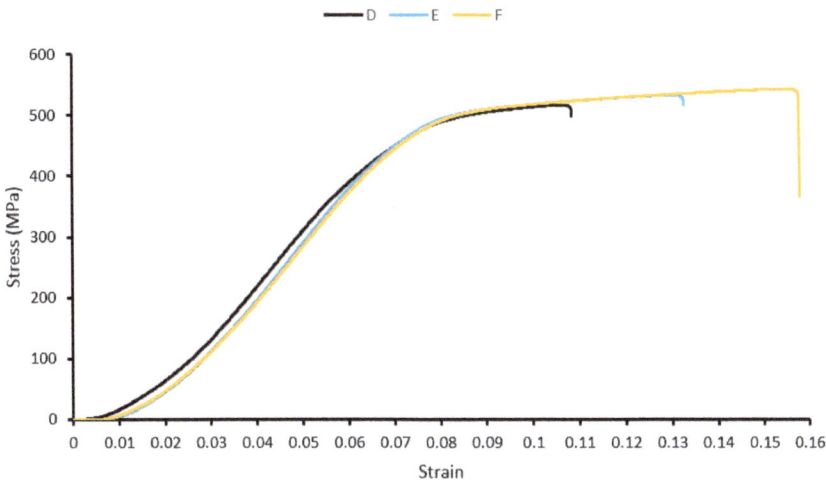

Figure 12. Customised specimen stress-strain plot.

4. Discussion

The results obtained for the porosity of the samples in Figure 4 clearly indicate an increase in porosity and pore size when progressing from parameter set A to C. The increased porosity may have arisen due to lack of fusion or Rayleigh instability, balling, or poor wetting characteristics as detailed by Rombouts, M. et al. [12]. With regards to the microstructure, long elongated grains were observed parallel to the building direction and consistent with other SLM work processing 316L [7]. These grains exhibit austenitic behaviour as is common in most 300 series stainless steels (316L included) [11,34]. The production of austenite grains is highly temperature and cooling rate dependent, with minimal variation observed in microstructures produced using samples sets A, B, and C, these observations are consistent with other work on SLM processing 316L with variable processing parameters [7].

From the stress-strain curve of the uniform sample set in Figure 7, it is clear that even though the energy density transferred to the sample was kept constant for samples A, B, and C, the change in parameters had an effect on the density of samples and subsequent mechanical properties. UTS and fracture strength have relatively close values, indicating signs of brittle behaviour, additionally supported/explained by the reduced necking at the specimen fracture point, results consistent with work conducted by Riemer, A., et al. [7]. A notable effect was observed in the 0.2% yield stress, UTS, and fracture stress, with a 29% decrease in yield, 33% decrease in UTS, and 38% decrease in fracture when progressing from parameter set A to C. There is a high variance in the measured maximum strain results with an average standard deviation of 19%. However, there is still a clear trend of decreasing max strain with decreasing laser power and increased exposure time, with a 49% decrease in average maximum strain from parameter set A to C. This decrease in average maximum strain can be attributed to the increase in porosity. According to ref. [35], decrease in porosity resulted in a significant improvement in elongation. Additionally, set A—having the larger maximum strain—exhibits more of a ductile behaviour by absorbing more energy per unit volume (area under stress-strain curve), and conversely by having a significantly lower maximum strain, set C exhibits a more brittle behaviour with less energy per unit of volume absorbed. In addition to these observations from the tensile test, the hardness testing results from Table 5 shows a trend of decreasing hardness with decreased laser power and increased exposure time, with a 21% decrease progressing from set A to C. Even with use of a constant energy density delivered to the SS 316L powder, variation in the process parameters leads to changes in the material behaviour.

For the customised specimens, samples fractured at the designed failure location which were created through processing specific layers with "weaker" processing parameters as in set C, while maintaining set A properties throughout the remainder of the sample. This introduces a predictable failure point within samples, and prevents randomness of fracture across the gauge length of samples (as those seen in uniform parameter set samples, Figure 6). The added capability of designing fracture locations within samples results using "hybrid" processing parameters (sets D, E, F) caused samples to have approximately a 2% lower yield strength and 7% lower UTS than standard uniforms samples (processed set A).

5. Conclusions

Whilst maintaining a consistent energy density, sample porosity varied as laser power and exposure time was modified. This change is linked to lack of fusion porosity or Rayleigh instabilities within the melt pool. It was observed that tensile samples produced with a uniform parameter set would fracture at random locations across its gauge length. Using a set of "hybrid" processing parameters, sample mechanical properties could be tailored such that specific fracture points within a sample could be designed into a component. Using two SLM processing parameters sets (hybrid parameters) to melt specific locations within a test piece, an additional 3% porosity across a segment making up approximately 1.7% of the sample's gauge length was generated. This was sufficient to initiate consistent and repeatable fracture across this segment while only reducing the yield strength of the entire sample by approximately 2% compared to uniform single set parameter specimens (these produced un-customised, random fracture locations).

This controlled in-built failure of components requires particular features to be "written" into the structure of a component through modification of SLM processing parameters at specific layers across a component. Controlled in-built failure of components and customisation of components' mechanical performance is a feature that is difficult to achieve with conventional metal forming techniques. Such customised components can form part of a larger assembly, which are intentionally engineered to exhibit particular behaviours when under specific external loading conditions in order to protect other components within the system. Examples of parts include rupture discs used for pressure relief in pressure vessels; blow-off panels used in enclosures, vehicles, or buildings where overpressure may occur; and shear pins preventing mechanical overloads. Using the approach of "hybrid" processing, parameters, and porosity within samples may be graded/adjusted layer to layer. This customisation of mechanical properties will exert more refined control over specific part fracture points or general mechanical behaviour in order to further enhance the overall performance or capabilities of a component. As research momentum in this underdeveloped area progresses, additive manufacturing will be able to lay claim and present an additional benefit of this technology. Further exploitation of the unique layer-by-layer building principle of SLM can lead to customisation of components' mechanical performance.

Author Contributions: Andrei Ilie planned experimentation and undertook majority of testing, Haider Ali assisted with testing, analysis and manuscript preparation and Kamran Mumtaz supervised overall project.

Conflicts of Interest: The authors declare no conflict of interest.

1. Childs, T.H.C.; Hauser, C.; Badrossamay, M. Mapping and Modelling Single Scan Track Formation in Direct Metal Selective Laser Melting. *CIRP Ann. Manuf. Technol.* **2004**, *53*, 191–194. [CrossRef]
2. Yadroitsev, I.; Bertrand, P.; Smurov, I. Parametric analysis of the selective laser melting process. *Appl. Surf. Sci.* **2007**, *253*, 8064–8069. [CrossRef]
3. Niendorf, T.; Leuders, S.; Riemer, A.; Richard, H.A.; Tröster, T.; Schwarze, D. Highly Anisotropic Steel Processed by Selective Laser Melting. *Metall. Mater. Trans. B* **2013**, *44*, 794–796. [CrossRef]
4. Yadroitsev, I.; Thivillon, L.; Bertrand, P.; Smurov, I. Strategy of manufacturing components with designed internal structure by selective laser melting of metallic powder. *Appl. Surf. Sci.* **2007**, *254*, 980–983. [CrossRef]
5. Fischer, P.; Romano, V.; Weber, H.P.; Karapatis, N.P.; Boillat, E.; Glardon, R. Sintering of commercially pure titanium powder with a Nd:YAG laser source. *Acta Mater.* **2003**, *51*, 1651–1662. [CrossRef]
6. Kruth, J.P.; Levy, G.; Klocke, F.; Childs, T.H. Consolidation phenomena in laser and powder-bed based layered manufacturing. *CIRP Ann. Manuf. Technol.* **2007**, *56*, 730–759. [CrossRef]
7. Riemer, A.; Leuders, S.; Thöne, M.; Richard, H.A.; Tröster, T.; Niendorf, T. On the fatigue crack growth behavior in 316L stainless steel manufactured by selective laser melting. *Eng. Fract. Mech.* **2014**, *120*, 15–25. [CrossRef]
8. Yasa, E.; Kruth, J.P. Microstructural investigation of Selective Laser Melting 316L stainless steel parts exposed to laser re-melting. *Procedia Eng.* **2011**, *19*, 389–395. [CrossRef]
9. Mertens, A.; Reginster, S.; Contrepois, Q.; Dormal, T.; Lemaire, O.; Lecomte-Beckers, J. Microstructures and mechanical properties of stainless steel AISI 316L processed by selective laser melting. *Mater. Sci. Forum* **2014**, *783*, 898–903. [CrossRef]
10. Antony, K.; Arivazhagan, N.; Senthilkumaran, K. Numerical and experimental investigations on laser melting of stainless steel 316L metal powders. *J. Manuf. Processes* **2014**, *16*, 345–355. [CrossRef]
11. Zhang, B.; Dembinski, L.; Coddet, C. The study of the laser parameters and environment variables effect on mechanical properties of high compact parts elaborated by selective laser melting 316L powder. *Mater. Sci. Eng. A* **2013**, *584*, 21–31. [CrossRef]
12. Rombouts, M.; Kruth, J.P.; Froyen, L.; Mercelis, P. Fundamentals of Selective Laser Melting of alloyed steel powders. *CIRP Ann. Manuf. Technol.* **2006**, *55*, 187–192. [CrossRef]
13. Simchi, A. Direct laser sintering of metal powders: Mechanism, kinetics and microstructural features. *Mater. Sci. Eng. A* **2006**, *428*, 148–158. [CrossRef]
14. Roberts, I.A. *Investigation of Residual Stresses in the Laser Melting of Metal Powders in Additive Layer Manufacturing*; University of Wolverhampton: Wolverhampton, UK, 2012; p. 246.

15. Knowles, C.R.; Becker, T.H.; Tait, R.B. The effect of heat treatment on the residual stress levels within direct metal laser sintered Ti-6Al-4V as measured using the hole-drilling strain gauge method. In Proceedings of the 13th international Rapid Product Development Association of South Africa (RAPDASA) Conference, Sun City, South Africa, 31 October–2 November 2012; pp. 1–10.

16. Chatterjee, A.N.; Kumar, S.; Saha, P.; Mishra, P.K.; Choudhury, A.R. An experiment design approach to selective laser sintering of low carbon steel. *J. Mater. Process. Technol.* **2003**, *136*, 151–157. [CrossRef]

17. Matsumoto, M.; Shiomi, M.; Osakada, K.; Abe, F. Finite element analysis of single layer forming on metallic powder bed in rapid prototyping by selective laser processing. *Int. J. Mach. Tools Manuf.* **2002**, *42*, 61–67. [CrossRef]

18. Kruth, J.P.; Deckers, J.; Yasa, E.; Wauthlé, R. Assessing and comparing influencing factors of residual stresses in selective laser melting using a novel analysis method. *Proc. Inst. Mech. Eng. Part B J. Eng. Manuf.* **2012**, *226*, 980–991. [CrossRef]

19. Joe Elambasseril, S.F.; Matthias, B.; Milan, B. *Influence of Process Parameters on Selective Laser Melting of Ti 6Al-4V Components*; RMIT University: Ho Chi Minh City, Vietnam, 2012.

20. Papadakis, L.; Loizou, A.; Risse, J.; Bremen, S. A thermo-mechanical modeling reduction approach for calculating shape distortion in SLM manufacturing for aero engine components. In Proceedings of the 6th International Conference on Advanced Research in Virtual and Rapid Prototyping, Leiria, Portugal, 1–5 October 2013.

21. Casavola, C.; Campanelli, S.L.; Pappalettere, C. Experimental analysis of residual stresses in the selective laser melting process. In Proceedings of the XIth International Congress and Exposition, Orlando, FL, USA, 2–5 June 2008.

22. Thöne, M.; Leuders, S.; Riemer, A.; Tröster, T.; Richard, H.A. Influence of heat-treatment on selective laser melting products–eg Ti6Al4V. In Proceedings of the Solid Freeform Fabrication Symposium SFF, Austin, TX, USA, 6–8 August 2012.

23. Shiomi, M.; Osakada, K.; Nakamura, K.; Yamashita, T.; Abe, F. Residual Stress within Metallic Model Made by Selective Laser Melting Process. *CIRP Ann. Manuf. Technol.* **2004**, *53*, 195–198. [CrossRef]

24. ASTM International. *Additive Manufacturing Technology Standards*; ASTM International: West Conshohocken, PA, USA, 2014.

25. Wohlers Associates, Inc. *Wohlers Report 2008: State of the Industry*; Annual Worldwide Progress Report; Wohlers Associates, Inc.: Fort Collins, CO, USA, 1996.

26. Gu, D.; Shen, Y. Balling phenomenon in direct laser sintering of stainless steel powder: Metallurgical mechanisms and control methods. *Mater. Des.* **2009**, *30*, 2903–2910. [CrossRef]

27. Yadroitsev, I.; Gusarov, A.; Yadroitsava, I.; Smurov, I. Single track formation in selective laser melting of metal powders. *J. Mater. Process. Technol.* **2010**, *210*, 1624–1631. [CrossRef]

28. Hanzl, P.; Zetek, M.; Bakša, T.; Kroupa, T. The Influence of Processing Parameters on the Mechanical Properties of SLM Parts. *Procedia Eng.* **2015**, *100*, 1405–1413. [CrossRef]

29. Vrancken, B.; Thijs, L.; Kruth, J.P.; Van Humbeeck, J. Heat treatment of Ti6Al4V produced by Selective Laser Melting: Microstructure and mechanical properties. *J. Alloys Compd.* **2012**, *541*, 177–185. [CrossRef]

30. ISO 6507-1:2005, Metallic Materials—Vickers Hardness Test—Test Method. Available online: http://www.iso.org/iso/catalogue_detail.htm?csnumber=37746 (accessed on 20 February 2017).

31. ASTM International. *Standard Test Methods for Determining Area Percentage Porosity in Thermal Sprayed Coatings*; E2109-01(2007); ASTM International: West Conshohocken, PA, USA, 2007.

32. British Standards Institution (BSI). *Method for Statistically Estimating the Volume Fraction of Phases and Constituents by Systematic Manual Point Counting with a Grid*; BS 7590:1992; BSI: London, UK, 1992.

33. ASTM International. *Standard Test Methods for Tension Testing of Metallic Materials*; E8/E8M-13a; ASTM International: West Conshohocken, PA, USA, 2013.

34. Reed-Hill, R.E.; Abbaschian, R. *Physical Metallurgy Principles*; PWS-Kent Publisher: Boston, MA, USA, 1992.

35. Qiu, C.; Adkins, N.J.E.; Attallah, M.M. Selective laser melting of Invar 36: Microstructure and properties. *Acta Mater.* **2016**, *103*, 382–395. [CrossRef]

technologies

MDPI

Article

Additive Manufacturing: Reproducibility of Metallic Parts

Konda Gokuldoss Prashanth [1,*], Sergio Scudino [2], Riddhi P. Chatterjee [2], Omar O. Salman [2] and Jürgen Eckert [1,3]

[1]   Erich Schmid Institute of Materials Science, Austrian Academy of Sciences, Jahnstraße 12,
    A-8700 Leoben, Austria; juergen.eckert@unileoben.ac.at
[2]   IFW Dresden, Institute for Complex Materials, Postfach 270116, D-01171 Dresden, Germany;
    s.scudino@ifw-dresden.de (S.S.); riddhipratim.chatterjee@gmail.com (R.P.C.);
    o.o.salman@ifw-dresden.de (O.O.S)
[3]   Department Materials Physics, Montanuniversität Leoben, Jahnstraße 12, A-8700 Leoben, Austria
*   Correspondence: kgprashanth@gmail.com or Prashanth.kondagokuldoss@oeaw.ac.at;
    Tel.: +43-3842-804-206; Fax: +43-3482-804-116

Academic Editors: Salvatore Brischetto, Paolo Maggiore, Carlo Giovanni Ferro and Manoj Gupta
Received: 20 December 2016; Accepted: 18 February 2017; Published: 22 February 2017

Abstract: The present study deals with the properties of five different metals/alloys (Al-12Si, Cu-10Sn and 316L—face centered cubic structure, CoCrMo and commercially pure Ti (CP-Ti)—hexagonal closed packed structure) fabricated by selective laser melting. The room temperature tensile properties of Al-12Si samples show good consistency in results within the experimental errors. Similar reproducible results were observed for sliding wear and corrosion experiments. The other metal/alloy systems also show repeatable tensile properties, with the tensile curves overlapping until the yield point. The curves may then follow the same path or show a marginal deviation (~10 MPa) until they reach the ultimate tensile strength and a negligible difference in ductility levels (of ~0.3%) is observed between the samples. The results show that selective laser melting is a reliable fabrication method to produce metallic materials with consistent and reproducible properties.

Keywords: selective laser melting; laser processing; metals and alloys; mechanical properties; tensile properties

1. Introduction

Ever since the manufacturing of materials took place in the Bronze Age, the existing techniques have been constantly developed and new manufacturing processes have been invented [1]. Conventional casting and powder metallurgy (powder production followed by consolidation) are two widely used manufacturing processes to produce parts for different applications [2–5]. Even though these processes are widely used, there are several problems associated with them. For example, the parts fabricated by conventional casting processes may tend to have one of the following processing defects: surface defects, internal defects, inconsistency in chemical composition (segregation) and/or unsatisfactory mechanical properties (inconsistencies in the grain structure) [6]. Similarly, the parts manufactured by powder metallurgy may have defects introduced at various stages during the fabrication chain, such as powder production (non-uniform chemical composition), powder compaction (porosity) and sintering (porosity and oxidation of surface) [7,8]. Defects can also originate during post-processing of the parts after fabrication [7]. All these defects, which are introduced at different stages of manufacturing or post-processing may lead to inferior/inconsistency in properties [9,10]. However, the stringent industrial regulations (automobile, aeronautical, power plants and nuclear industry) nowadays require parts to have highly reproducible mechanical properties [11]. To conform to the stringent regulations, efforts have been made to find alternative processing routes or to reduce the unreliability factor in the existing processing capabilities. Additive manufacturing is seen as one of the viable alternative processing routes which may lead to consistent properties in materials. The laser-based powder bed fusion process (ISO/ASTM52900:2015 Standard Terminology for Additive Manufacturing–General Principles–Terminology), which is commonly known as Selective Laser Melting (SLM) is one of the additive manufacturing processes, which produces three-dimensional metal parts layer by layer with superior properties compared to conventional manufacturing processes such as casting and powder metallurgy [12–15]. A suitable combination of the processing parameters such as the laser power, laser scan speed, hatch distance, hatch style, layer thickness and laser spot size leads to the fabrication of a defect-free component by SLM [12]. The above-mentioned parameters, with the exception of hatch style, determine the heat/energy supplied to the powder bed (heat/energy input). The amount of powder surface exposed to the laser during the SLM process is rather small and hence a very high energy density is involved during the SLM process. This intense energy input leads to very high cooling rates observed in the rate of $\sim 10^5$–10^6 K/s [16,17]. Such high cooling rates will result in substantially refined microstructures compared to the conventional manufacturing processes, and hence improved properties [16,17].

The majority of the research on the SLM process is focused on parameter optimization, alloy development, topology optimization/structure optimization and microstructure-property correlation. The intent of the present manuscript is to highlight the repeatability/reproducibility aspects of the samples produced by SLM in terms of the material properties. Five different materials—Al-12Si, Cu-10Sn, and 316L, belonging to the face-centered cubic structure, CoCrMo and CP-Ti with hexagonal closed packed structure—were evaluated and their properties are reported to show consistency in the properties of the samples produced by SLM.

2. Experimental Section

Cylindrical tensile samples (total length 52 mm, length and diameter of the gauge length 17.5 and 3.5 mm) were fabricated from spherical gas-atomized powders at room temperature using an SLM 250 HL device (from SLM Solutions and formerly Machine Tool Technologies Solutions). The device is equipped with a Yb-YAG laser. All the samples were built over a base plate made of the same material as the building material under an Ar environment (in order to avoid oxygen contamination during the building process) with a hatch style rotation of 73°. Hatch style is defined as the design/pattern in which the hatches (melting sequences or melt lines or melt tracks) are oriented within and between the layers [18]. Detailed information about the hatch style can be found in [19]. All the samples were built perpendicular to the base plate (i.e., *XY* direction). An allowance of 1–2 mm was given for these samples, so that they can be machined with the abrasive papers to smoothen their surface before the tensile test. The tensile test samples used in the present study were selected randomly from different batches at randomly built positions in the substrate plate, in order to ascertain the reproducibility criteria and were used in the as-built condition. The Al-12Si samples (from gas atomized powder with a nominal composition of Al-12Si (wt.%)) were fabricated with a laser power of 320 W for both the bulk of the sample and the contour and the laser scan speed of 1455 mm/s for the bulk and 1939 mm/s for the contour. A layer thickness of 50 μm is used with a laser spot size of ~80 μm and a hatch distance of ~110 μm. Detailed information about the fabrication of the Al-12Si samples can be found elsewhere [20]. The following parameters were used for the fabrication of CoCrMo parts (from CoCrMo gas atomized powder from SLM solutions): laser power—100 W; laser scan speed—140 mm/sec; layer thickness—30 μm; and hatch distance—100 μm, with 90° hatch rotation between the layers [21]. For further processing details about CoCrMo, see [21]. Commercially pure Ti (CP-Ti) samples were built from CP-Ti grade 2 powder supplied by TLS Technik GmbH, Germany with the following parameters: laser power—165 W; laser scan speed—138 mm/s; layer thickness—100 μm; and hatch distance—100 μm, with 90° hatch rotation between the layers. Detailed information about the fabrication of the CP-Ti can be found at [22]. Gas atomized 316L powders were used to fabricate SLM parts with the following parameters: laser power—100 W; laser scan speed—800 mm/s; layer thickness—30 μm; and hatch distance—120 μm, with 90° hatch rotation between the layers [23,24]. Similar gas atomized bronze powders (with the following parameters: laser power—271 W; laser scan speed—210 mm/s; laser thickness—90 μm; and hatch distance—90 μm) were used for producing the bulk SLM bronze parts [25]. Cylindrical bulk samples were also prepared by graphite mold casting in order to compare the properties of the conventionally fabricated cast samples with the SLM samples. Room temperature tensile tests were carried out using an Instron 8562 testing facility (strain rate 1×10^{-4} s^{-1}) and the strain during the tensile test was measured directly on the specimen using a Fiedler laser-extensometer. At least three specimens were tested under each condition to ascertain the reproducibility/repeatability of the properties. The wear and corrosion test conditions have been reported elsewhere [26]. For the corrosion experiments, the samples were mounted in polymer resin and are polished metallographically to mirror finish. A Solarton SI Electrochemical Interface connected to a tempered three-electrode-cell with a Pt net as the counter electrode was used, with a saturated calomel reference electrode (SCE with Standard Hydrogen Electrode potential E_{SHE} = 0.241 V at room temperature) and the embedded alloy sample as the working electrode. Before the actual polarization measurements, the samples were kept at open circuit potential (OCP) conditions for 1 h; meanwhile, the potential was monitored. The linear dynamic polarization was started at −0.2 V vs. OCP, and the potential was increased at a constant rate of 0.5 mV/s up to a value of 1.5 V vs. SCE.

3. Results and Discussion

Figure 1 shows the room temperature tensile curves of the Al-12Si samples manufactured by SLM and the corresponding mechanical data are summarized in Table 1. The tensile curves (in color) in Figure 1a show the consolidated data of six tensile tests that are shown individually in Figure 1b. It is observed from Figure 1a that all the six tensile curves almost overlap one another and only negligible differences are observed after yielding and at the time of fracture. The yield strength (YS) of these six samples varies between 239 and 242 MPa and the ultimate tensile strength (UTS) varies between 375 and 384 MPa. The ductility of the sample varies between 2.65% and 2.85%, showing consistency in the tensile properties. Zhang et al. and Li et al. have also shown that tensile properties of Al-12Si samples prepared by SLM under Ar atmosphere lie in the range: YS—235 and 250 MPa; UTS—370 and 390 MPa; and ductility 2.75% and 3% [27,28]. Interestingly, both Zhang et al. and Li et al. fabricated the Al-12Si samples with the SLM device from ReaLizer and not from SLM solutions. Siddique et al. has also produced Al-12Si samples by SLM using the SLM solutions device and the tensile properties lie within the above said range [29]. This suggests that with the optimized parameters for full density, the Al-12Si SLM samples will show repeatable/reproducible tensile properties within the experimental errors. However, there are reports of anisotropy in the SLM produced samples and the mechanical properties vary depending on the building direction [30–33]. Alsalla et al. have shown that the tensile strength and the fracture toughness of the 316L cellular lattice manufactured by the SLM technique, depends greatly on the building direction. This is essentially due to the anisotropic behavior of the SLM-prepared samples [30]. Similar anisotropy has been reported by Suryawanshi et al., where the fracture toughness of the Al-12Si samples depends strongly on the building direction [17]. On the other hand, some results also suggest that the sample building direction does not have a significant effect on the tensile properties [20]. Hence, there exists a contradiction between the consistencies in the tensile properties of samples prepared with different build orientation. However, it may be safe to say that even if there is a difference in properties between the samples prepared with different build directions, the differences are consistent and reproducible within the experimental limits. This suggests that the samples built in each orientation (XY/YZ/XZ) should give repeatable and reproducible mechanical properties, when tested in similar conditions.

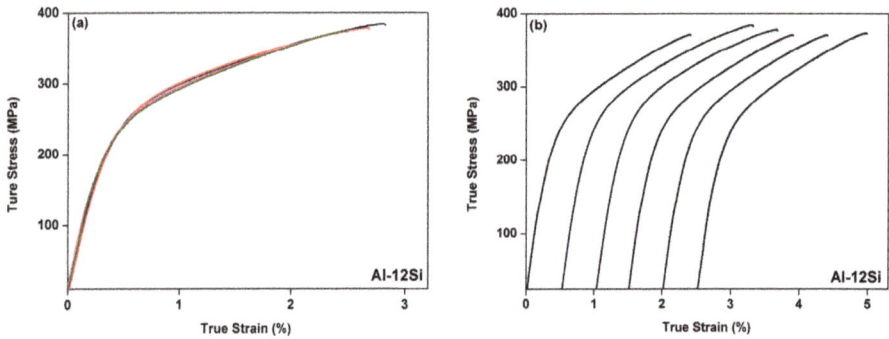

Figure 1. Room temperature tensile test curves for the Al-12Si SLM samples (**a**) consolidated data; (**b**) individual tensile curves.

Table 1. Tensile properties of samples produced by selective laser melting (SLM) and casting (cast).

Properties \ Samples Designation	Al-12Si— SLM	Al-12Si— Cast	316L— SLM	316L— Cast	CoCrMo— SLM	CoCrMo— Cast
Yield Strength (MPa)	240 ± 1	62 ± 9	495 ± 3	254 ± 45	764 ± 2	621 ± 22
Ultimate Tensile Strength (MPa)	380 ± 4	166 ± 48	836 ± 7	573 ± 56	1201 ± 10	908 ± 55
Fracture strain (%)	2.8 ± 0.1	10 ± 4	35.0 ± 0.5	18 ± 6	12.7 ± 0.5	5.8 ± 2

The sliding wear test data of Al-12Si SLM samples are shown in Figure 2a. The data points (corresponding to sample number 1) are the consolidated data points of six sliding wear test experiments that are shown as samples 2–7 (Figure 2a). The wear test results are quite repeatable with the wear rate varying between 9.23 and 9.24 × 10^{-13} m^3/m, showing consistency within the experimental errors. Similar results have been observed for the corrosion studies (conducted in an acidic HNO$_3$ medium), where the potentiodynamic polarization curves between two test samples almost overlap each other except for small but negligible differences within the experimental limits (Figure 2b). The above results indicate that the tensile properties, wear rate and potentiodynamic corrosion results obtained for the Al-12Si samples produced by SLM are very consistent and reproducible in nature. It might be thought that the Al-12Si samples show consistent and reproducible properties because both Al and Si phases constituting the structure have a face centered cubic (fcc) crystal structure. Hence, to further check the reproducibility of the mechanical properties of SLM parts, other fcc systems such as Cu-10Sn bronze and 316L (predominantly austenite phase) and hexagonally closed packed (hcp) systems, CoCrMo and commercially pure Ti (CP-Ti), were evaluated.

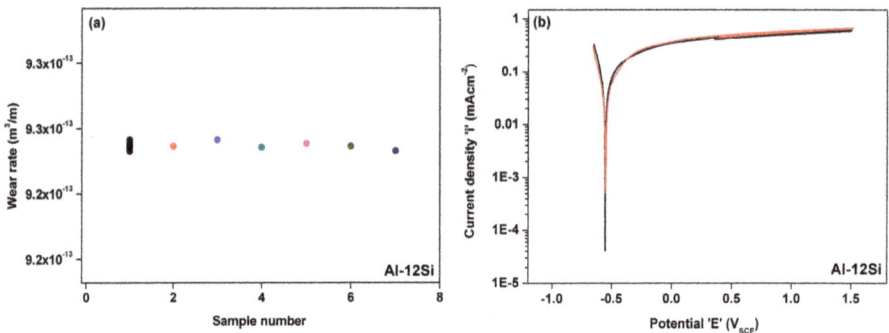

Figure 2. (**a**) Wear rate data for Al-12Si samples carried out at 10 N load; (**b**) Potentiodynamic polarization curves of Al-12Si SLM samples measured in acidic 1 M HNO$_3$ (pH = 0).

Figure 3 shows the room temperature tensile tests for CoCrMo, 316L, commercially pure Ti (CP-Ti) and Cu-10Sn bronze alloys. Two tensile curves for each alloy are shown in a consolidated fashion (in color) followed by their individual tensile curves (in black). The consolidated curves for Cu-10Sn overlap and no significant differences are found from the tensile test results. A similar trend is observed for the 316L samples, where a marginal difference of ~8 MPa in YS is realized between two tensile tests along with a difference in UTS of ~15 MPa and ductility of ~0.3%. The tensile test curves for CP-Ti do not show any difference in YS between two tensile test results and a marginal difference in UTS and ductility of ~5 MPa and ~0.3%, respectively, is observed. Similar results were found in the case of the SLM-processed CoCrMo alloy. The alloy shows a difference of ~1 MPa in YS between two tensile test results and the difference between the UTS and ductility is ~9 MPa and ~0.45%, respectively. The tensile properties of the Al-12Si, 316L and CoCrMo samples fabricated by casting are shown in Table 1. It can be observed that the samples fabricated by casting have inferior strengths. Moreover, the cast samples show a larger standard deviation compared to the samples fabricated by SLM. The above results from different alloy systems reveal that the SLM-processed materials show very good consistency in their properties (mechanical, tribological and corrosion properties) within the experimental errors, even though the samples were picked randomly from several batches (8–10 batches over 1 year in the case of Al-12Si). The placement of the sample during the building process was also selected randomly. The results were conclusive that the sample batches, irrespective of the sample position, will yield similar, consistent and reproducible properties, if the hardware remains the same along with the quality of the laser. This is because the same hardware with the same quality of laser source will yield a similar amount of defects (porosity level) and hence similar or reproducible properties. This suggests that the SLM process can lead to the production of metals and alloys with superior as well as more reproducible properties compared to their counterparts produced by conventional casting.

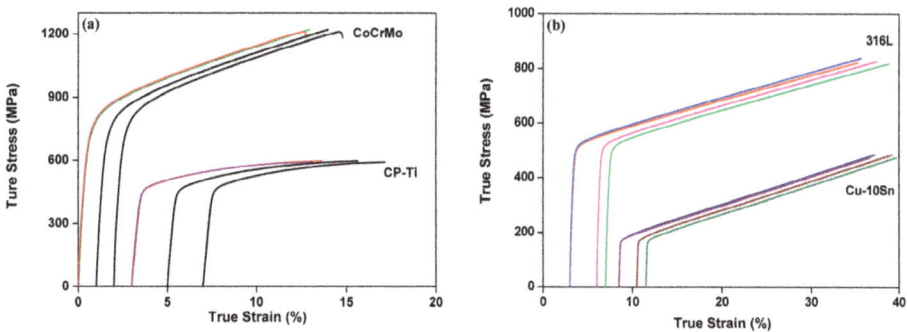

Figure 3. Room temperature tensile curves for (**a**) CoCrMo and commercially pure Ti (CP-Ti) and (**b**) 316L and Cu-10Sn bronze materials.

4. Conclusions

Five different metal/alloy systems (Al-12Si, Cu-10Sn and 316L—face centered cubic phase and CoCrMo and CP-Ti—hexagonal closed packed phase) were fabricated by SLM using commercially available parameters. The Al-12Si fcc samples show uniform and consistent mechanical, tribological and corrosion properties within the experimental errors. It is noteworthy that the room temperature tensile curves overlap one another up to the yield point and show similar behavior, beyond yielding or marginal differences in the ultimate tensile strength (difference ~10 MPa) and/or ductility (~0.2%), thus demonstrating the reliability of the samples fabricated by SLM. Similar tensile results were observed in the case of the other four metal/alloy systems (Cu-10Sn, 316L, CoCrMo and CP-Ti), where the room temperature curves show consistency in their mechanical properties. These results suggest that the selective laser melting process can be used to produce parts with consistent and reproducible properties, provided the powder quality and the parameters for fabrication remain the same.

Acknowledgments: The authors would like to thank A. Gebert for providing the facilities to carry out the corrosion experiments and for stimulating technical discussions. Financial support through the German Science Foundation (DFG) under the Leibniz Program (grant EC 111/26-1) and the European Research Council under the ERC Advanced Grant INTELHYB (grant ERC-2013-ADG-340025) is gratefully acknowledged.

Author Contributions: Konda Gokuldoss Prashanth, Sergio Scudino, and Jürgen Eckert conceived and designed the experiments; Konda Gokuldoss Prashanth performed the experiments on Al-12Si, Ti and Cu-10Sn alloys; Omar O. Salman performed the experiments on 316L and Riddhi P. Chatterjee on CoCrMo alloys. Konda Gokuldoss Prashanth and Sergio Scudino analyzed the data; all the authors have contributed to the discussion of the results and the writing of the paper.

Conflicts of Interest: The authors declare no conflict of interest.

1. Historical Development of Materials and Manufacturing Process. 2016. Available online: http://me-mechanicalengineering.com/historical-development-of-materials-and-manufacturing-process/ (accessed on 15 June 2016).
2. Groover, M.P. *Introduction to Manufacturing Processes*; John Wiley & Sons: New York, NY, USA, 2012.
3. Shercliff, H.R.; Lovatt, A.M. Selection of manufacturing processes in design and the role of process modelling. *Prog. Mater. Sci.* **2001**, *46*, 429–459. [CrossRef]
4. Wang, Z.; Prashanth, K.G.; Chaubey, A.K.; Löber, L.; Schimansky, F.P.; Pyczak, F.; Zhang, W.W.; Scudino, S.; Eckert, J. Tensile properties of Al-12Si matrix composites reinforced with Ti-Al based particles. *J. Alloys Compd.* **2015**, *630*, 256–259. [CrossRef]
5. Wang, Z.; Prashanth, K.G.; Scudino, S.; Chaubey, A.K.; Sordelet, D.J.; Zhang, W.W.; Li, Y.Y.; Ecker, J. Tensile properties of Al matrix composites reinforced with in situ devitrified $Al_{84}Gd_6Ni_7Co_3$ glassy particles. *J. Alloys Compd.* **2014**, *586*, S419–S422. [CrossRef]
6. Rajkolhe, R.; Khan, J.G. Defects, causes and their remedies in casting process: A review. *Int. J. Res. Advent Technol.* **2014**, *2*, 375–383.
7. Gutmanas, E.Y. Materials with fine microstructures by advanced powder metallurgy. *Prog. Mater. Sci.* **1990**, *34*, 261–366. [CrossRef]
8. Khan, M.I.; Haque, S. *Manufacturing Science*; PHI Learning Private Limited: New Delhi, India, 2011.
9. Wang, H.T.; Fang, Z.G.Z.; Sun, P. A critical review of mechanical properties of powder metallurgy titanium. *Int. J. Powder Metall.* **2010**, *46*, 45–57.
10. Klar, E.; Samal, P. *Powder Metallurgy Stainless Steels: Processing, Microstructures and Properties*; ASM International: Materials Park, Ohio, USA, 2007.
11. Pandiripalli, B. *Repeatability and Reproducibility Studies: A Comparison of Techniques*; Wichita State University: Wichita, KS, USA, 2010.
12. Olakanmi, E.O.; Cochrane, R.F.; Dalgarno, K.W. A review on selective laser sintering/melting (SLS/SLM) of aluminium alloy powders: Processing, microstructure and mechanical properties. *Prog. Mater. Sci.* **2015**, *74*, 401–477. [CrossRef]
13. Prashanth, K.G.; Shahabi, H.S.; Attar, H.; Srivastava, V.C.; Ellendt, N.; Uhlenwinkel, V.; Eckert, J.; Scudino, S. Production of high strength $Al_{85}Nd_8Ni_5Co_2$ alloy by selective laser melting. *Addit. Manuf.* **2016**, *6*, 1–5. [CrossRef]
14. Prashanth, K.G.; Scudino, S.; Chaubey, A.K.; Löber, L.; Wang, P.; Attar, H.; Schimansky, F.P.; Pyczak, F.; Eckert, J. Processing of Al-12Si-TNM composites by selective laser melting and evaluation of compressive and wear properties. *J. Mater. Res.* **2016**, *31*, 55–65. [CrossRef]
15. Sun, Z.; Tan, X.; Tor, S.B.; Yeong, W.Y. Selective laser melting of stainless steel 316L with low porosity and high build rates. *Mater. Des.* **2016**, *104*, 197–204. [CrossRef]
16. Pauly, S.; Löber, L.; Petters, R.; Stoica, M.; Scudino, S.; Kühn, U.; Eckert, J. Processing metallic glasses by selective laser melting. *Mater. Today* **2013**, *16*, 37–41. [CrossRef]

17. Suryawanshi, J.; Prashanth, K.G.; Scudino, S.; Eckert, J.; Prakash, O.; Ramamurty, U. Simultaneous enhancements of strength and toughness in an Al-12Si alloy synthesized using selective laser melting. *Acta Mater.* **2016**, *115*, 285–294. [CrossRef]
18. Prashanth, K.G.; Scudino, S.; Eckert, J. Defining the tensile properties of Al-12Si parts produced by selective laser melting. *Acta Mater.* **2017**, *126*, 25–35. [CrossRef]
19. Prashanth, K.G. Selective Laser Melting of Al-12Si. Ph.D. Thesis, Technische Universität Dresden, Dresden, Germany, 2014.
20. Prashanth, K.G.; Scudino, S.; Klauss, H.J.; Surreddi, K.B.; Löber, L.; Wang, Z.; Chaubey, A.K.; Kühn, U.; Eckert, J. Microstructure and mechanical properties of Al-12Si produced by selective laser melting: Effect of heat treatment. *Mater. Sci. Eng. A* **2014**, *590*, 153–160. [CrossRef]
21. Chatterjee, R.P. Selective Laser Melting of Co-Cr-Mo Alloys. Master of Technology Thesis, Indian Institute of Technology Kharagpur, West Bengal, India, 2015.
22. Attar, H.; Prashanth, K.G.; Chaubey, A.K.; Calin, M.; Zhang, L.C.; Scudino, S. Comparison of wear properties of commercially pure titanium prepared by selective laser melting and casting processes. *Mater. Lett.* **2015**, *142*, 38–41. [CrossRef]
23. Löber, L. Selektives Laserstrahlschmelzen von Titanaluminiden und Stahl. PhD Thesis, Technische Universität Dresden, Dresden, Germany, 2015.
24. Prashanth, K.G.; Löber, L.; Klauss, H.; Kühn, U.; Eckert, J. Characterization of 316L steel cellular dodecahedron structures produced by selective laser melting. *Technologies* **2016**, *4*, 34. [CrossRef]
25. Scudino, S.; Unterdörfer, C.; Prashanth, K.G.; Attar, H.; Ellendt, N.; Uhlenwinkel, V.; Eckert, J. Additive manufacturing of Cu-10Sn bronze. *Mater. Lett.* **2015**, *156*, 202–204. [CrossRef]
26. Prashanth, K.G.; Debalina, B.; Wang, Z.; Gostin, P.F.; Gebert, A.; Calin, M.; Kühn, U.; Kamaraj, M.; Scudino, S.; Eckert, J. Tribological and corrosion properties of Al-12Si produced by selective laser melting. *J. Mater. Res.* **2014**, *29*, 2044–2054. [CrossRef]
27. Wang, X.J.; Zhang, L.C.; Fang, M.H.; Sercombe, T.B. The effect of atmosphere on the structure and properties of a selective laser melted Al-12Si alloy. *Mater. Sci. Eng. A* **2014**, *597*, 370–375. [CrossRef]
28. Li, X.P.; Wang, X.J.; Saunders, M.; Suvorova, A.; Zhang, L.C.; Liu, Y.J.; Fang, M.H.; Huang, Z.H.; Sercombe, T.B. A selective laser melting and solution heat treatment refined Al-12Si alloy with a controllable ultrafine eutectic microstructure and 25% tensile ductility. *Acta Mater.* **2015**, *95*, 74–82. [CrossRef]
29. Siddique, S.; Imran, M.; Wycisk, E.; Emmelmann, C.; Walther, F. Influence of process-induced microstructure and imperfections on mechanical properties of AlSi12 processed by selective laser melting. *J. Mater. Process. Technol.* **2015**, *221*, 205–213. [CrossRef]
30. Alsalla, H.; Hao, L.; Smith, C. Fracture toughness and tensile strength of 316L stainless steel cellular lattice structures manufactured using the selective laser melting technique. *Mater. Sci. Eng. A* **2016**, *669*, 1–6. [CrossRef]
31. Kajima, Y.; Takaichi, A.; Nakamoto, T.; Kimura, T.; Yogo, Y.; Ashida, M.; Doi, H.; Nomura, N.; Takahashi, H.; Hanawa, T.; et al. Fatigue strength of Co-Cr-Mo alloy clasps prepared by selective laser melting. *J. Mech. Behav. Biomed. Mater.* **2016**, *59*, 446–458. [CrossRef] [PubMed]
32. Tang, M.; Pistorius, P.C. Oxides, porosity and fatigue performance of AlSi$_{10}$Mg parts produced by selective laser melting. *Int. J. Fatigue* **2017**, *94*, 192–201. [CrossRef]
33. Rosenthal, I.; Stern, A.; Frage, N. Strain rate sensitivity and fracture mechanism of AlSi$_{10}$Mg parts produced by selective laser melting. *Mater. Sci. Eng. A* **2017**, *682*, 509–517. [CrossRef]

technologies

MDPI

Article
Emergence of Home Manufacturing in the Developed World: Return on Investment for Open-Source 3-D Printers

Emily E. Petersen [1] and Joshua Pearce [1,2,*]

[1] Department of Material Science and Engineering, Michigan Technological University, Houghton, MI 49931, USA; eepeters@mtu.edu
[2] Department of Electrical and Computer Engineering, Michigan Technological University, Houghton, MI 49931, USA
*   Correspondence: pearce@mtu.edu; Tel.: +1-906-487-1466

Academic Editors: Salvatore Brischetto, Paolo Maggiore and Carlo Giovanni Ferro
Received: 13 January 2017; Accepted: 6 February 2017; Published: 9 February 2017

Abstract: Through reduced 3-D printer cost, increased usability, and greater material selection, additive manufacturing has transitioned from business manufacturing to the average prosumer. This study serves as a representative model for the potential future of 3-D printing in the average American household by employing a printer operator who was relatively unfamiliar with 3-D printing and the 3-D design files of common items normally purchased by the average consumer. Twenty-six items were printed in thermoplastic and a cost analysis was performed through comparison to comparable, commercially available products at a low and high price range. When compared to the low-cost items, investment in a 3-D printer represented a return of investment of over 100% in five years. The simple payback time for the high-cost comparison was less than 6 months, and produced a 986% return. Thus, fully-assembled commercial open source 3-D printers can be highly profitable investments for American consumers. Finally, as a preliminary gauge of the effect that widespread prosumer use of 3-D printing might have on the economy, savings were calculated based on the items' download rates from open repositories. Results indicate that printing these selected items have already saved prosumers over $4 million by substituting for purchases.

Keywords: distributed manufacturing; additive manufacturing; 3-D printing; consumer; economics; open-source

1. Introduction

Private manufacturing, also referred to as household manufacturing, has a lengthy history in the United States which resulted in the emergence of domestic commerce [1,2]. With the development of interchangeable parts, however, came the assembly line, and manufacturing transitioned to standardized high-volume mass production [3,4]. Lower variable costs, greater flexibility, and higher average product performance contributed significantly to this transition [5]. Since then, a global trend toward large-scale, centralized manufacturing and global shipping, particularly for inexpensive plastic products, has arisen alongside growing world consumerism [6,7]. Economies of scale provided consumers with more convenient and lower-priced goods than what they could make themselves [8]. However, the rapid growth of the 3-D printing industry may change this trend.

Additive manufacturing (AM), or 3-D printing, promises to be an emerging 21st century innovation platform for promoting distributed manufacturing for many products [9–13]. The compound annual growth rate of worldwide additive manufacturing products and services over the past three years, from 2013 to 2015, was 31.5% [14]. Although a less centralized model of manufacturing than that currently practiced, the conventional 3-D printing industry is still focused on businesses manufacturing and selling products to consumers or other businesses [14]. However, with the rise of Internet sharing and open source hardware development [15], it may provide a more aggressive path to distributed production. Most notably, the self-replicating rapid prototyper (RepRap) 3-D printer [16–18] can fabricate more than half of its own parts. Already, RepRaps have significantly reduced distributed digital manufacturing costs for high-end products such as scientific equipment and have enabled economic non-business distributed manufacturing [19–21]. The savings for the distributed manufacturing of these high-end products [22] provide staggering value for the scientific community [23,24]. However, distributed manufacturing is not relegated to high-price specialty items.

Preliminary research has already shown that the number of free pre-designed 3-D products is growing rapidly, and low-cost do-it-yourself (DIY) 3-D printers such as the RepRap are already economically beneficial for the average American consumer [25]. This provides the opportunity for the most radical form of distributed manufacturing. At-home 3-D printing capitalizes on the elimination of product transport, establishing the technology within the realm of distributive manufacturing's three-tiered modes of operation [25] (tier 1: central manufacturing distributed to different locations, tier 2: decentralized further to local and agile production sites (e.g., localized manufacturing , fablabs, and makerspaces), and tier 3: at home manufacturing). Nonetheless, in order for this innovative form of localized and customized manufacturing to make a significant impact on the industry as a whole, ease of use and the economic advantage to the average consumer must be better understood [26]. In particular, the past study by Wittbrodt et al. [25] assumed that the consumer was technically savvy enough to build a 3-D printer from parts using freely available Internet plans. This may have been an overly optimistic assumption as less than a third of Americans are scientifically and technically literate [27,28]. Considering past work in the context of the technical sophistication of the American public, two questions arise: Will 3-D printing be relegated largely to replacing conventional manufacturing techniques and creating the potential for more distributed business-based manufacturing [29,30]? Alternatively, can 3-D printing be used to economically manufacture in the majority of American homes of technically illiterate people? In addition, it is worth acknowledging that financial savings provide just one contribution to a consumer's motivations, so economic analysis must be kept in context.

To probe this latter question of the economic viability of this scale of 3-D printing for home manufacturing in the developed world, this study reports on the life-cycle economic analysis (LCEA) of Lulzbot Mini technology for an average U.S. household. The Lulzbot Mini is a commercialized and fully assembled plug-and-play derivative of the RepRap, which can be used by a consumer with no training and modest technical familiarity [31]. A selection of twenty-six freely available open-source 3-D printable designs that a typical first-world household might purchase were selected to simulate use over half a year at the average rate of production of one "home-made" item per week. A selection of the parts was printed to determine energy use per mass of material. Printed products were quantified by print time and filament consumption by mass and the experimental masses and printing time were compared to slicer software estimates. The experimental values were converted to the cost to the user and were then compared to low and high market prices for comparable commercially available products. The results of this life-cycle economic analysis provide a return on investment (ROI) for the prosumer (producing consumer), which is compared to other potential investments. Finally, the downloaded substitution value of the selection of designs is quantified to draw conclusions about the future of manufacturing in developed-world economies.

2. Materials and Methods

For this analysis, it was critical that the methods of manufacturing and materials were relevant and accessible to the average consumer. A Lulzbot Mini [31] was selected due to the ease of use, high resolution capabilities, support of open-source hardware and software, and the ability to work with a variety of operating systems, as well as its relevance in the 3-D printing community following other similar products [32]. To be used by the Mini, 3-mm poly lactic acid (PLA) was selected as the filament because it is the most common household printing material. PLA has gained prominence, as not only does it demonstrate less warping during printing than other materials such as the second most common 3-D printing plastic (ABS), but the emissions during printing are less pungent [32,33]. Furthermore, PLA is made from corn-based resin, making it non-toxic, biodegradable, and able to be produced in environmentally friendly, renewable processes [34,35]. It should be noted that because the ABS filament costs are roughly equivalent to PLA and the melting temperature is not that much higher, the results from this study can be extrapolated to ABS.

Twenty-six items were selected from open source 3-D printable design repositories after searching for open source design files indexed on Yeggi.com, which is a 3-D design file search engine. The twenty-six items are summarized in the Supplementary Materials including the source of the design, and the low and high price URLs for roughly equivalent products. Items were selected to represent the average American consumer's use over the course of half a year of printing one product per week. Three criteria were used in the selection of products: (1) printable by a Lulzbot Mini in PLA (e.g., having an appropriate build volume, resolution, and material requirements); (2) widely considered to be a common product purchased (or class of product purchased) or owned by the average American consumer; and (3) has a commercially comparable alternative available for purchase online. The concluding analysis was mindful of the difficulty in quantifying the print quality, however the items included in this study met the authors' expectations for acceptable quality (e.g., z-level print lines are observable using the high-quality quick print settings, but not unacceptable for general consumer use).

One of the most challenging areas in 3-D printing technical knowledge for new users is optimizing the slicer settings that determine the tool path of the 3-D printer. To avoid this challenge, all parts were printed in PLA using the QuickPrint settings in the Lulzbot version of Cura [36] to demonstrate ease of use. Figure 1 shows the Lulzbot Mini mid-print using PLA and Cura Quick Print settings. The estimated and actual mass, filament length, and estimated and actual printing time were recorded. All parts were weighed on an electronic balance with an error of ± 0.02 g.

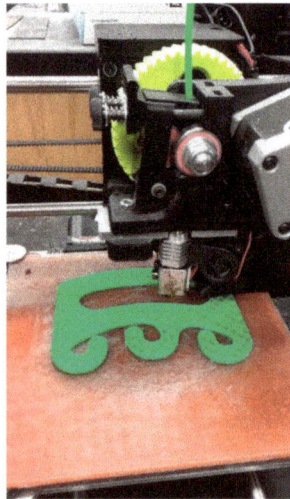

Figure 1. Lulzbot Mini mid-print using poly lactic acid and Cura Quick Print settings.

In order to apply a cost per hour for each printed item, the print time and energy consumption was recorded by a multimeter (± 0.02 kWh) for complex, simple, and average geometric complexity. Greater print complexity demonstrated a higher level of energy consumption primarily because of the operation time per unit mass. The average was found to be about 0.01 kWh/g, which is higher than that reported in the Wittbrodt et al. study [25] due to the additional energy consumed by the heated bed of the Mini. The average consumption of 0.01 kWh/g was applied for all prints included in the study.

High and low commercial prices for each product was found primarily on Walmart.com and supplemented using Google Shopping. Associated shipping costs were excluded from the analysis for both purchasing and distributed manufacturing (e.g., no shipping charges included for the plastic filament). The operating cost for the Lulzbot Mini (O_L) was calculated using the electricity and filament consumption during printing. The average electricity rate in 2015 in the United States is $0.1267 per kWh for the residential sector [37] and the cost of a 1 kg spool of 3-mm PLA was found to range between $23/kg and $25/kg [38,39], so $24/kg was used here. This operating cost was calculated as follows:

$$O_L = EC_E + \frac{C_F m_f}{1000} \text{ [USD]} \tag{1}$$

where E is the energy consumed in kWh, C_E is the average rate of electricity in the United States in USD/kWh, C_F is the average cost of the PLA filament in USD/kg, and m_f is the mass of the filament in grams consumed during printing. Thus, the total cost (C_T) to the average consumer using the selected printer an average of once per week is the following:

$$C_T = \sum_0^T O_L + C_L \text{ [USD]} \tag{2}$$

where the operating cost is summed over T years and C_L is the cost of the Lulzbot Mini itself. It should be noted here that the capital costs were not considered because it is assumed that the prosumer is not financing the cost of the capital equipment because the Lulzbot Mini cost is only $1250.00 [40]. It should be pointed out that if the 3-D printer were purchased on a credit card, which is the only feasible method of financing a consumer purchase such as this, this would need to be included in C_L. This would also be true if the prosumer purchased a large amount of inventory filament on credit. C_T was evaluated over a range of years from one to five. The marginal savings on each project, Cs, is given by:

$$C_S = C_C - O_L \text{ [USD]} \tag{3}$$

where C_C is the cost of the commercially available product (which is calculated for both low and high online prices), and the marginal percent change, P, between the cost to print a product and the commercially available product was calculated as follows:

$$P = \frac{C_C - O_L}{C_C} \times 100 \text{ [\%]} \tag{4}$$

where C_C is the cost for the commercial product at either the high or low price.

When the cost of the 3-D printer is taken into account the total savings, S, is given by:

$$S = \sum_0^T \sum_0^A C_C - C_T \text{ [USD]} \tag{5}$$

over T years and all, A, products.

The simple payback time of the printer (t_{pb}) was calculated by the following:

$$t_{pb} = \frac{C_L}{S} \text{ [years]} \tag{6}$$

An estimated return on investment (R) was calculated following [41,42] assuming a five-year lifetime for the 3-D printer.

$$\frac{C_L}{S} = \frac{1 - e^{-RT}}{R} \qquad (7)$$

Finally, the value obtained from a free and open source 3-D printable design can be determined from the downloaded substitution valuation, $V_D(t)$ [23,24] at a specific time (t):

$$V_D(t) = (C_p - C_m) \times p \times N_d(t) \text{ [USD]} \qquad (8)$$

This value is determined by the number of downloads (N_d) on 7 December 2016, where Cp is the retail cost of the traditionally manufactured product and C_m is the marginal cost to fabricate it with the Lulzbot mini. p is the percent of downloads resulting in a print. It should be noted that p is subject to error as downloading a design does not guarantee its manufacture. On the other more likely hand, a single download could be fabricated many times, traded via email, memory stick, or posted on P2P websites that are beyond conventional tracking. Here, to remain conservative, p is assumed to be 1 because downloading a design involves effort that is not repaid unless one does the printing. This is equivalent to assuming that if a consumer downloads an ebook that it is read at least once. It is thus reasonable to assume every download resulted in at least one print and the total savings for the random 26 objects can be conservatively determined by:

$$V_{DT}(t) = \sum_{i=0}^{26} V_D(t) \text{ [USD]} \qquad (9)$$

All economic values are in U.S. dollars (USD), $.

3. Results

Printing twenty-six items to model use over the course of 6 months resulted in a total of 104.18 m of filament consumption with a mass of 737.8 g. An estimated total of 100.18 h, and 7.26 kWh were expended on 3-D printing. This translates to $17.71 worth of material and $0.92 in electricity based on average U.S. electric rates, for a total operational cost of $18.63 over half a year. Thus, at a printing rate of one object per week, operating the 3-D printer costs less than $40 per year. Table 1 shows the projected cumulative cost of owning and using a Lulzbot Mini as a function of years.

Table 1. The projected cumulative cost for owning a Lulzbot Mini increased from year 1 to 5. The cost includes the price of the printer itself, the cost of the filament, and the energy consumption to print an average product per week.

Year	Cumulative Cost of Ownership (USD)
1	1287
2	1325
3	1362
4	1399
5	1436

Retail costs for the products totaled $278.57 and $1376.03 for low- and high-priced items, respectively, as seen in Table 2.

Table 2. 3-D printable object, length of filament, print time, energy consumption, low and high retail price, operating cost, marginal savings, and percent savings for 26 freely available designs.

Object	Length (m)	Weight (g)	Total Print Time (min)	Energy Consumption (kWh)	Price: Low (USD)	Price: High (USD)	Operating Cost (USD)	Marginal Savings: Low (USD)	Marginal Savings: High (USD)	Marginal Percent Change: Low	Marginal Percent Change: High
Spoon holder	1.55	10.69	139	0.11	7.75	29.00	0.2701	7.48	28.73	96.51	99.07
Arduino nano enclosure	2.84	20.44	168	0.20	8.82	9.95	0.5165	8.30	9.43	94.41	94.81
Carpet corner	0.15	0.84	8	0.01	9.99	10.30	0.0212	9.97	10.28	99.79	99.79
Bathroom wine glass holder	5.75	41.01	359	0.41	8.70	44.96	1.0362	7.66	43.92	88.09	97.70
Tool holder	1.65	11.75	106	0.12	4.70	21.10	0.2969	4.40	20.80	93.68	98.59
Soap holder	7.60	54.57	482	0.55	4.86	12.99	1.3789	3.48	11.61	71.63	89.39
Snowboard bind plate	4.51	30.54	265	0.31	11.95	12.96	0.7716	11.18	12.19	93.54	94.05
Dremel cutting table	10.06	67.13	619	0.67	26.00	43.99	1.6961	24.30	42.29	93.48	96.14
Rotary tool attachment	2.12	14.16	152	0.14	12.98	49.99	0.3578	12.62	49.63	97.24	99.28
Solder stand	4.24	29.31	276	0.29	13.65	28.95	0.7406	12.91	28.21	94.57	97.44
Nikon lens cap holder	2.04	14.26	118	0.14	5.15	7.05	0.3673	4.79	6.69	93.00	94.89
Speaker grill	2.84	18.73	173	0.19	9.70	50.95	0.4733	9.23	50.48	95.12	99.07
Espresso tamper	5.48	36.77	334	0.37	17.86	40.99	0.9291	16.93	40.06	94.80	97.73
Sewing machine presser foot	0.24	1.48	21	0.01	5.99	20.49	0.0374	5.95	20.45	99.38	99.82
Coin holder	8.30	58.23	540	0.58	9.88	64.87	1.4713	8.41	63.40	85.11	97.73
Shower head	10.49	68.11	183	0.68	5.18	110.61	1.7209	3.46	108.89	66.78	98.44
Seatbelt guide	0.79	4.82	53	0.05	15.00	27.99	0.1218	14.88	27.87	99.19	99.56
Trumpet mute	3.93	26.53	379	0.27	11.95	99.99	0.6703	11.28	99.32	94.39	99.33
Ski pole GoPro mount	1.08	7.39	74	0.07	7.96	99.99	0.1867	7.77	99.80	97.65	99.07
Canon lens hood	1.61	9.25	116	0.09	5.99	52.99	0.2337	5.76	52.76	96.10	99.56
Insulin belt clip	0.69	3.85	51	0.04	15.99	22.41	0.0908	15.89	22.31	99.39	99.57
Torque wrench nozzle	2.59	21.02	165	0.24	4.20	419.58	0.5349	3.67	419.05	87.26	99.87
Rodin figurine	6.98	54.9	380	0.53	22.32	46.99	1.3847	20.94	45.61	93.80	97.05
iPhone6 case	2.37	19.10	100	0.13	0.99	59.99	0.4749	0.52	59.52	52.03	99.91
Deathstar model	4.77	38.02	285	0.45	25.98	35.00	0.9695	25.01	34.03	96.27	97.23
Pokemon planter	9.51	74.90	465	0.61	5.03	31.95	1.8749	3.16	30.08	62.73	97.13
Total	104.18	737.80	6011	7.26	278.57	1376.03	18.63	259.94	1357.40	Average 93.31	Average 98.65

This results in a substantial prosumer savings for each individual product with an average marginal cost reduction of 93.3% and 98.7% when compared against the low and high retail costs, respectively. This results in total savings of $259.94 and $1357.40 for the low and high cost estimates, respectively. Table 3 shows the projected prosumer profit per year assuming one product fabrication per week using the average of the 26 objects chosen here when compared to low-cost commercially available products and high-priced commercially available products. As shown, profit is realized after the second year of ownership when only low-cost commercially available products are considered in the analysis. When compared to high-cost products, however, profit is realized within the first year of ownership. It should be noted that as many of the objects allow some form of customization, the latter values are a better estimate for comparison.

Table 3. The profit to the user was projected when printed items, produced at a rate of one per week, were compared to low and high cost comparable, commercially available products.

Year	Low-Cost Projected Profit (USD)	High-Cost Projected Profit (USD)
1	−730.11	1464.81
2	−210.23	4179.61
3	309.66	6894.42
4	829.55	9609.23
5	1349.43	12,324.03

Comparing the printed objects to the lowest-priced equivalent product, there was a payback time of 2.4 years. In comparison to high-priced items, payback time was only 0.46 years. The return on investment was 25% in year 3 and 108% by year 5 when low-range cost values were considered. Comparing printing costs to high-end commercial prices resulted in a 552% ROI in year 3 and 986% in year 5.

The number of downloads for each item file was used to estimate the total savings for the global 3-D printing community when compared to marginal savings using the high and low commercial prices. These values are shown in Table 4. When compared to the low-end prices, the 26 printed items saved $803,945.70. Compared to the high-end prices, the savings were $4,033,657.89. The URLs for the designs and the low/high equivalent products are found in Supplementary Material. Prior research has provided economic justification for quantifying these projected values [23].

Table 4. The downloaded substitution value for the 26 free design example files.

Object	Number of Downloads	Low Marginal Savings of Downloads (USD)	High Marginal Savings of Downloads (USD)
Spoon holder	3113	23,248.92	89,436.17
Arduino nano enclosure	157	1303.66	1481.07
Carpet corner support	382	3808.07	3926.49
Bathroom wine glass holder	847	6491.24	37,203.46
Tool holder	534	2356.60	11,108.86
Soap holder	492	1712.74	5712.70
Snowboard bind plate	154	1801.80	1957.34
Dremel cutting table	7027	170,782.99	297,198.72
Rotary tool attachment	6080	76,743.09	301,763.89
Solder stand	1765	22,785.15	49,789.63
Nikon lens cap holder	1312	6284.08	8776.88
Speaker grill	233	2149.83	77,761.08
Espresso tamper	1195	20,232.46	47,872.81
Sewing machine presser foot	235	1398.86	4806.36
Coin holder	183	1538.79	11,601.96
Shower head	3921	42,691.94	369,272.03
Seatbelt guide	119	1708.50	3254.31

Table 4. *Cont.*

Object	Number of Downloads	Low Marginal Savings of Downloads (USD)	High Marginal Savings of Downloads (USD)
Trumpet mute	561	6585.59	55,976.03
Ski pole GoPro mount	121	873.34	2328.97
Canon lens hood	3421	20,135.60	180,922.60
Insulin belt clip	383	6044.91	8503.77
Torque wrench nozzle	3797	13,918.25	1,591,116.11
Rodin figurine	8129	170,163.13	370,705.56
iPhone6 case	1175	608.10	69,933.10
Deathstar model	6829	170,800.02	232,397.60
Pokemon planter	8807	37,742.05	264,850.38
Totals		**803,945.70**	**4,033,657.89**

4. Discussion

The impact of introducing additive manufacturing to the average American home demonstrates both microeconomic and macroeconomic advantages.

4.1. Microeconomic Advantages of Home Manufacturing

With projected savings to a single owner per year of $519.89 and $2,714.81 when compared to low and high range commercially available items respectively, the Lulzbot Mini could serve as a significant means by which the average consumer can reduce personal expenses. The items selected for the study represent those frequently found in the home such as tool mounts, shower heads, seat belt guides, figurines, and espresso tampers. From the perception of the consumer, they can begin to substitute free designs and 3-D printed objects for high-end consumer purchases such that printing only a single product a week recovers the cost of the printer in under a year. Some prosumers will use their 3-D printers considerably more than that and will be able to recover their initial investment more rapidly by printing out the same types of items at a greater rate or in a few expensive substitutional prints (e.g., custom orthotics) [25].

It is instructive to consider the purchase of a consumer-friendly printer as an investment and compare it to more traditional investments available to the average consumer. For example, five-year CD rates have 1.85%–2.10% APY [43] and savings account rates on investments less than $5000 and even those over $100,000 (jumbo) only go up to 1.05% [44]. In the volatile stock market, the historic corporate earnings have gone up an average of 7% per year. Thus by comparison, the return on investment demonstrated here with distributed manufacturing in the home of common products is an extremely positive outlook for the average consumer. When compared to low-range commercial prices, the return on investment was over 100% within five years of ownership. In comparison to high-priced items, the return was a staggering 117% by the end of the first year. Within three years, the return grew to 552%. It should again be pointed out that all estimates for the purpose of this study remained conservative, but a consumer's willingness to accept the perceived risk of such an investment is based largely on their discount rate, not a comparison to their other available investment options.

Discount rate, a frequent point of contention in the literature, has been confirmed to vary among consumers depending on factors including income, race, and education [45]. For instance, with increasing consumer education the discount rates used for decision-making decreases [45]. High, triple-digit discount rates have been used as some studies have attempted to determine "implicit consumer discount rates" [45–49]. The lack of information and consumer illiteracy regarding alternative investments (e.g., energy consumption information) has contributed to the greater trend of the un-educated and poorly educated making unfavorable economic decisions, thus lending to higher observed discount rates [49]. Previous studies such as [46] and modern studies have erroneously argued for the application of implicit discount rates (e.g., 27% to 102%) to low and median-income households, reserving low discount rates for the efficiency standards for high-income households [50].

The unattainability of three-digit returns on low-risk investment opportunities to the lower and middle class highlight the implausibility of such policy recommendations, and if adopted would advance the ignorant dialogue of economic errors commonly observed in the American lower and middle class. By quantifying the time value for money and risk associated with future cash flow, it is possible to establish a model of discount rates. In the closest investment analog, the U.S. Department of Energy (DOE) has looked at consumers treating home energy conservation measures as investments. The DOE has established a set of energy efficiency standards for common household appliances, and in order to economically justify investment in reducing electrical consumption, a sensitivity analysis was performed. The discount rate in the study was conservative, varying from 3% to 7% [51]. A 3% "social rate of time preference" was approximated to mirror the average saving rate using the real rate of return on long-term government debt [52]. In this way, a model can be established for how American consumers value current and future consumption. The 7% limit was set as the marginal rate of return on an average stock market investment prior to taxes [51]. Thus, in general 1% to 7% should be used in the sensitivity analysis by determining the amount of printing a prosumer would need to do at the average savings per print found in this study to reach a 1% and 7% return to break even. As can be seen by the results, the ROI from distributed manufacturing with 3-D printers surpasses these rates by orders of magnitude and are even competitive with the implicit consumer discount rates. This indicates that sales of prosumer 3-D printers will continue to climb as discussed in the next section.

It should also be pointed out that the return on a 3-D printer investment by a consumer is all tax free as they represent a reduction in consumer spending. In addition, consumers would reduce their personal taxes in a second way as they would only pay taxes on the investment of the 3-D printer and the filament. Thus they would avoid the sales taxes on all substituted products. These savings along with the savings on shipping were not included here, but would only assist in driving the ROIs for the purchase of a 3-D printer higher for an individual consumer.

4.2. Macroeconomic Advantages of Home Manufacturing

Despite the relatively low 0.7% growth of the United States economy in the final quarter of 2015, consumer spending remained steady due to a steady gain in jobs and rising wages [53,54]. Consumers remained cautious in 2015 as personal saving rates reached at or near their highest levels since 2012, laying the groundwork for economically beneficially and innovative technology to penetrate the at-home consumer market [55]. This trend contributed to projected consumer trends in 2016 including the automated creation of ideas and designs, resource sharing, and personalization [56,57]. Accessible additive manufacturing with new lower-cost 3-D printers, such as the Lulzbot Mini used as an example in this study, fit into this trend, providing the average consumer with an economic alternative to commercial purchasing and a platform through which to innovate and collaborate with other users. The high return on investment values calculated from this study show a clear advantage to the average consumer, even when compared to low-priced commercial alternatives.

Furthermore, the transition of additive manufacturing from industry to the consumer market has followed the growing trend of conscious consumerism [58]. By providing a means by which to make products, consumers develop a heightened level of responsibility and become more selective in their consumerism [59]. In addition, it is clear that distributed AM represents an environmental benefit because of reduced material use, transportation, and the elimination of packaging [60–62], and a growing contingent of responsible consumers are considering environmental concerns into their purchase decisions [63–67]. This has encouraged a more vibrant do-it-yourself (DIY) community, one that is driven not only by saving money but also by the enjoyment of the experience [68]. DIY production implies a negative impact on government tax income, which needs to be investigated in more detail in the future. In addition, there could be an impact on employment/unemployment rates through its substitution within industrial production/increase in at-home businesses, which needs to be further investigated. Early analysis [69] saw at-home additive manufacturing's niche use for customizable, small, high-value items. However, 3-D printing can be used for far more than such a limited range of products as shown in this study. Low-cost 3-D printers have enabled emerging additive manufacturing technology to transition from industry and academia to the average consumer, resulting in a market that has exploded from 66 purchased printing units in 2007 to 23,265 units in 2011 [70]. Improvements have made this technology both technically accessible and economically advantageous to the consumer market [25].

Significant savings at the macroscopic level appears to already be occurring for early-adopting prosumers. The savings of just the 26 printed items used as examples in this study have already saved consumers over $4 million when the number of downloads recorded on Thingiverse.com and Youmagine.com are combined and compared to high priced retail goods. As many designers post their files on other depository sites in addition to free and open access pages, these estimates are again conservative values. It is noteworthy that these two websites have over 2 million free and open source designs, while dozens of other repositories exist [71]. The values found in this paper indicate that distributed manufacturing by prosumers could have a substantial economic impact in the near future as the number of 3-D printer users and free designs continue to climb. Furthermore, the items selected for the purpose of this study were not placed in the greater context of item popularity which would further increase the download rate and thus the projected savings. Finally, it should be noted that the items were all freely accessible having been created by a global network of makers and shared under open licenses. In the analysis, the operation cost per minute (filament plus electricity) was calculated to total $0.08/min for all 26 items. This did not include personnel costs as would normally be calculated for a business manufacturing an item. The user time is truly limited as operation requires a "time investment" equivalent to approximately the cost of time for online shopping thanks to pre-made designs (e.g., instead of inputting credit card information, prosumers download the stl and click print). Thus, here it was ignored because when using the quick print settings, as soon as the stl is loaded, the user clicks print and can then walk away and has no active participation in the manufacturing. In general it takes less than 1 minute to load an stl, have it slice in Cura, and click print (it should noted, that large complex designs take longer to slice). Thus for this study, it can be assumed that roughly half an hour of user time was invested and thus the prosumer's hourly rate for making their own products was over $500/h to over $2,600/h based on low and high value product estimations, respectively. In some cases, (e.g., the last two designs) it appears to be the 3-D equivalent of 'fan art'. The reality of the ease with which this is done challenges both the premise of patent law [72] as well as the viability of current intellectual property laws covering trademarks and copyright. Significant future work is needed to determine how to optimize the use of the concepts of intellectual property to maximize the public benefit.

4.3. Limitations and Future Work

There are several areas of technical study that would improve the viability of distributed manufacturing with 3-D printers including: (1) materials selection; (2) reliability; and (3) first costs.

First, the range of materials would expand the potential list of products that can be substituted with 3-D printing. Material selection has contributed to the freedom that 3-D printing presents to the average user, but it is far from complete. Not only does a variety of materials available to the average consumer exist, but the environmentally-friendly nature of a filament such as PLA is in line with the transition of consumer markets to green consumerism [73]. This trend could be further supported by adopting new polymer recycling codes to further expand the materials selection while reducing costs without introducing otherwise negative environmental impacts [74]. Furthermore, the lifetime of 3-D printed products is a topic of future work, as negative perceptions of low-lifetime prints could inhibit adoption within the greater manufacturing community.

The reliability of prosumer 3-D printers can be lower than experiences consumers are accustomed to with more mature products. The most common failure mode in fused filament fabrication (FFF) 3-D printing is nozzle clogging during printing due to one or a combination of the following mechanisms:

particulate contamination from the printing environment, contamination on the exterior or the interior of the filament, and non-uniform properties of the material within the extruder. In addition, older filament can be brittle and break before entering the extruder, ruining the print. These errors represent catastrophic failures during a print, some of which can be after several hours of printing. Wittbrodt et al. estimated that such errors represented 20% of all prints for new users with a self-built RepRap 3-D printer [25]. Such errors are significantly reduced for systems like the Lulzbot Mini; however, they still exist. To correct that error here, a systematic approach toward troubleshooting was adopted and can be used by inexperienced consumers. First, the filament was removed on heating to extract contaminants from the head and the extrusion head was cleaned. The filament end was cut so that a clean edge was used upon re-insertion. This level of maintenance is possible for most prosumers, but the uptake among consumers can be expected to increase if such tasks are automated/eliminated in the future. As long as well-designed 3-D printable parts are chosen, the error rate is ~0%; however, novice printer users are likely to choose some designs that are not conducive to perfect FFF style printing and then these higher errors should be taken into account. Future work could involve a detailed study of many novice printer users for actual behavior and printable part selection.

In this study it was assumed that the prosumer simply purchased the relatively low-cost printer and filament with cash. As noted in the methods these values would change if purchased on credit. It is not anticipated that the average consumer would utilize credit card financing in order to purchase a printer; however, considering the average purchase interest rate is 12.51% on all accounts and 13.76% on interest-bearing accounts [75], these interest rates are dwarfed by the ROI of 3-D printing products at home as found in the results (e.g., 25% in year 3 and 108% by year 5 when low-range cost values were considered and 552% ROI in year 3 and 986% in year 5).

Finally, although prosumer 3-D printers can easily pay for themselves by printing a modest number of consumer products, prices for most prosumer 3-D printers are still greater than $1000. This price makes them more expensive than the average laptop computer, which limits their market. In addition, there are other motivations to current consumption patterns (e.g., some consumers have been trained by marketers to consider shopping a leisure activity, conspicuous consumption provides positive peer feedback in some demographic groups, and fitting into U.S. consumer culture), which may impact their willingness to adopt distributed manufacturing despite overwhelming economic benefits.

Future work is needed to quantify the downloaded substitution value on all of the currently available free 3-D printable designs along with the possible savings for new types of commercially available materials such as flexible polymers. These studies could be better supported with surveys of users to develop a more refined value of *p* and a more accurate knowledge of how prosumers utilize their 3-D printers (e.g., rate of use, printing available designs vs. making their own, etc.). In addition, future work could analyze consumers' willingness to purchase a 3-D printer by comparing rational economic savings as was done here to large iconic personal prints for cultural status.

5. Conclusions

This study shows a clear financial advantage to owning and using prosumer-friendly printers. Additive manufacturing has demonstrated a clear advantage in reducing the cost of research equipment, however penetration into consumer markets has proven to be more difficult due to printer usability and print qualities. By employing a printer operator who was relatively unfamiliar to 3-D printing and printing files considered common items used by the average American consumer, this study serves as a representative model for the potential future of 3-D printing in the home. With a calculated return on investment of over 100% within three and one year of ownership when compared to low and high price ranges respectively, this study has shown that 3-D printers are an economically advantageous purchase for the average consumer. In addition, based on the downloaded substitution value of the 26 example products used here already being over $4 million, there is an indication of significant macroeconomic impact in the future as more consumers purchase 3-D printers and use them to fabricate freely available digital designs to offset conventional product purchases.

Supplementary Materials: The following are available online at www.mdpi.com/2227-7080/5/1/7/s1, Table S1: Sources of design files and low and high retail prices.

Acknowledgments: All the authors would like to thank Emeka Esemonu, Harris Kenny, and Ben Wittbrodt for helpful discussions and the support from Aleph Objects.

Author Contributions: Joshua Pearce conceived and designed the experiments; Emily E. Petersen performed the experiments; Joshua Pearce and Emily E. Petersen analyzed the data; Joshua Pearce contributed materials/analysis tools; and Joshua Pearce and Emily E. Petersen wrote the paper. Both authors have read and approved the final manuscript.

Conflicts of Interest: The authors have no conflicts of interest.

1. Tryon, R.M. *Household Manufactures in the United States 1640–1860: A Study in Industrial History*; The University of Chicago Press: Chicago, IL, USA, 1917.

2. Sokoloff, K.; Villaflor, G. The Market for Manufacturing Workers. In *The Market for Manufacturing Workers during Early Industrialization: The American Northeast, 1820 to 1860*; Goldin, C., Rockoff, H., Eds.; University of Chicago Press: Chicago, IL, USA, 1992.

3. Hounshell, D. *From American System to Mass Production, 1800–1932*; Johns Hopkins University Press: Baltimore, MD, USA, 1984.

4. Fine, C.; Freund, R. *Economic Analysis of Product-Flexible Manufacturing System Investment Decisions*; Massachusetts Institute of Technology: Cambridge, MA, USA, 1986.

5. Wilson, J. Henry Ford vs. Assembly Line Balancing. *Int. J. Prod. Res.* **2013**, *52*, 757–765. [CrossRef]

6. Kravis, I.; Lipsey, R. *Towards an Explanation of National Price Levels*; Working paper series 1034; National Bureau of Economic Research: Cambridge, MA, USA, 1982.

7. Lipsey, R. *Challenges to Home- and Host-Country Effects of Foreign Direct Investment: Analyzing the Economics, Challenges to Globalization: Analyzing the Economics*; Baldwin, R.E., Winters, A., Eds.; University of Chicago Press: Chicago, IL, USA, 2004.

8. Bain, J. Economies of Scale, Concentration and the Condition of Entry in Twenty Manufacturing Industries. *Am. Econ. Rev.* **1954**, *44*, 15–39.

9. Scan, B. How to Make (almost) Anything. The Economist. Available online: http://www.economist.com/node/4031304 (accessed on 6 February 2017).

10. Gershenfeld, N. How to Make Almost Anything: The Digital Fabrication Revolution. Available online: http://cba.mit.edu/docs/papers/12.09.FA.pdf (accessed on 6 February 2017).

11. Markillie, P. A Third Industrial Revolution. The Economist. Available online: http://www.economist.com/node/21552901 (accessed on 6 February 2017).

12. Gwamuri, J.; Wittbrodt, B.; Anzalone, N.; Pearce, J. Reversing the Trend of Large Scale and Centralization in Manufacturing: The Case of Distributed Manufacturing of Customizable 3-D-Printable Self-Adjustable Glasses. *Chall. Sustain.* **2014**, *2*, 30–40. [CrossRef]

13. Wittbrodt, B.; Laureto, J.; Tymrak, B.; Pearce, J. Distributed Manufacturing with 3-D Printing: A Case Study of Recreational Vehicle Solar Photovoltaic Mounting Systems. *J. Frugal Innov.* **2015**, *1*, 1–7. [CrossRef]

14. Wohlers Associates Inc. *Wohlers Report 2016: 3D Printing and Additive Manufacturing State of the Industry Annual Worldwide Progress Report*; Wohlers Associates Inc.: Fort Collins, CO, USA, 2016.

15. Gibb, A.; Abadie, S. *Building Open Source Hardware: DIY Manufacturing for Hackers and Makers*, 1st ed.; Addison-Wesley Professional: Boston, MA, USA, 2014.

16. Sells, E.; Bailard, S.; Smith, Z.; Bowyer, A.; Olliver, V. RepRap: The Replicating Rapid Prototyper-Maximizing Customizability by Breeding the Means of Production. In Proceedings in the World Conference on Mass Customization and Personalization, Cambridge, MA, USA, 7–10 October 2007.

17. Jones, R.; Haufe, P.; Sells, E.; Iravani, P.; Olliver, V.; Palmer, C.; Bowyer, A. RepRap-the Replicating Rapid Prototyper. *Robotica* **2011**, *29*, 177–191. [CrossRef]

18. Bowyer, A. 3D Printing and Humanity's First Imperfect Replicator. *3D Print. Addit. Manuf.* **2014**, *1*, 4–5. [CrossRef]

19. Pearce, J. Building Research Equipment with Free, Open-Source Hardware. *Science* **2012**, *337*, 1303–1304. [CrossRef] [PubMed]

20. Pearce, J. *Open-Source Lab: How to Build Your Own Hardware and Reduce Research Costs*, 1st ed.; Elsevier: Waltham, MA, USA, 2014.
21. Baden, T.; Chagas, A.; Marzullo, T.; Prieto-Godino, L.; Euler, T. Open Laware: 3-D Printing Your Own Lab Equipment. *PLoS Biol.* **2015**, *13*, e1002086.
22. Blua, A. A New Industrial Revolution: The Brave New World of 3D Printing. Radio Free Europe/Radio Liberty. 2013. Available online: http://www.rferl.org/a/printing-3d-new-industrial-revolution/24949765. html (accessed on 6 February 2017).
23. Pearce, J. Quantifying the Value of Open Source Hardware Development. *Mod. Econ.* **2015**, *6*, 1–11. [CrossRef]
24. Pearce, J.M. Return on investment for open source scientific hardware development. *Sci. Public Policy* **2016**, *43*, 192–195. [CrossRef]
25. Wittbrodt, B.; Glover, A.; Laureto, J.; Anzalone, G.; Oppliger, D.; Irwin, J.; Pearce, J. Life-Cycle Economic Analysis of Distributed Manufacturing with Open-Source 3-D Printers. *Mechatronics* **2013**, *23*, 716–726. [CrossRef]
26. Gilpin, L. 3D Printing: 10 Factors Still Holding It Back. TechRepublic. 2014. Available online: http://www.techrepublic.com/article/3d-printing-10-factors-still-holding-it-back/ (accessed on 6 February 2017).
27. Duncan, D. 216 Million Americans Are Scientifically Illiterate (Part I). MIT Technology Review 2007. Available online: https://www.technologyreview.com/s/407346/216-million-americans-are-scientifically-illiterate-part-i/ (accessed on 6 February 2017).
28. Miller, J. The Public Understanding of Science in Europe and the United States. In Proceedings in the American Association for the Advancement of Science, San Francisco, CA, USA, 16 February 2007.
29. Laplume, A.; Petersen, B.; Pearce, J. Global value chains from a 3D printing perspective. *J. Int. Bus. Stud.* **2016**, *47*, 595–609. [CrossRef]
30. Laplume, A.; Anzalone, G.; Pearce, J. Open-source, self-replicating 3-D printer factory for small-business manufacturing. *Int. J. Adv. Manuf. Technol.* **2015**, *85*, 633–642. [CrossRef]
31. Hoffman, T. LulzBot Mini 3D Printer. PCMAG. 2015. Available online: http://www.pcmag.com/article2/0, 2817,2476575,00.asp (accessed on 6 February 2017).
32. Hoffman, T. LulzBot TAZ 5 3D Printer. PCMAG. 2015. Available online: http://www.pcmag.com/article2/ 0,2817,2489833,00.asp (accessed on 6 February 2017).
33. Chilson, L. The Difference between ABS and PLA for 3D Printing. *ProtoParadigm.* 2013. Available online: http://www.protoparadigm.com/news-updates/the-difference-between-abs-and-pla-for-3d-printing/ (accessed on 6 February 2017).
34. Stephens, B.; Azimi, P.; El Orch, Z.; Ramos, T. Ultrafine Particle Emissions from Desktop 3D Printers. *Atmos. Environ.* **2013**, *79*, 334–339. [CrossRef]
35. Tokiwa, Y.; Calabia, B.; Ugwu, C.; Aiba, S. Biodegradability of Plastics. *Int. J. Mol. Sci.* **2009**, *10*, 3722–3742. [CrossRef] [PubMed]
36. Ultimaker. Cura and 3D Printing Made for Each Other. 2016. Available online: https://ultimaker.com/en/ products/cura-software (accessed on 6 February 2017).
37. U.S. Energy Information Administration Independent Short-Term Energy Outlook (STEO). Available online: https://www.eia.gov/forecasts/steo/pdf/steo_full.pdf (accessed on 6 February 2017).
38. Amazon Hatchbox 3D PLA-1kg 3.00-BLK PLA 3D Printer filament Dimensional Accuracy +/− 0.05 Mm, 1 Kg Spool, 3.00 Mm, Black. Available online: http://www.amazon.com/HATCHBOX-3D-PLA-1KG3-00-BLK-Filament-Dimensional/dp/B00MEZE7XU (accessed on 6 February 2017).
39. Lulzbot PLA. Available online: https://www.lulzbot.com/store/filament/pla-esun (accessed on 6 February 2017).
40. LulzBot Mini. Available online: https://www.lulzbot.com/store/printers/lulzbot-mini (accessed on 6 February 2017).
41. Pearce, J.M.; Denkenberger, D.; Zielonka, H. Accelerating applied sustainability by utilizing return on investment for energy conservation measures. *Int. J. Energy Environ. Econ.* **2009**, *17*, 61–80.
42. Pearce, J.M.; Denkenberger, D.; Zielonka, H. Accelerating Applied Sustainability by Utilizing Return on Investment for Energy Conservation Measures. *Adv. Energy Res.* **2011**, *3*, 107–126.
43. Bankrate CD Rates. Available online: http://www.bankrate.com/funnel/cd-investments/cd-investment-results.aspx?prods=19&market=1324 (accessed on 12 September 2016).

44. Bestrate Savings and Money Market Accounts. Available online: http://accounts.bestrates.com/savings-and-money-market-accounts (accessed on 12 September 2016).

45. Newell, R.; Siikamaki, J. Individual Time Preferences Energy Efficiency. *Am. Econ. Rev.* **2015**, *105*, 196–200. [CrossRef]

46. Hausman, J. Individual Discount Rates and the Purchase and Utilization of Energy Using Durables. *Bell J. Econ.* **1979**, *10*, 33–54. [CrossRef]

47. Dermot, G. Individual Discount Rates and the Purchase and Utlization of Energy Using Durables. *Bell J. Econ.* **1980**, *10*, 33054.

48. Ruderman, H.; Levine, M.; McMahon, J. The Behavior of the Market for Energy Efficiency in Residential Appliances Including Heating and Cooling Equipment. *Energy J.* **1987**, *8*, 101–124. [CrossRef]

49. Frederick, S.; Lowenstein, G.; O'Donoghue, T. Time Discounting and Time Preference: A Critical Review. *J. Econ. Lit.* **2002**, *40*, 351–401. [CrossRef]

50. Miller, S. One Discount Rate Fits All? The Regressive Effects of DOE's Energy Efficiency Rule. *Policy Perspect.* **2015**, *22*, 40–54. [CrossRef]

51. OMB (Office of Management and Budget). 1992. Circular A-94. Guidelines and Discount Rates for Benefit-Cost Analysis of Federal Programs.

52. OMB (Office of Management and Budget). 2003. Circular A-94. Guidelines and Discount Rates for Benefit-Cost Analysis of Federal Programs.

53. Torrey, H. Economists React to Fourth-Quarter U.S. Growth: 'Consumer Spending and Housing Remain Solid.' Newpaper. *Wall Street Journal Blogs*, 2016.

54. Trading Economics United States Consumer Spending: 1950–2016. Available online: http://www.tradingeconomics.com/united-states/consumer-spending (accessed on 12 September 2016).

55. Sparshott, J. U.S. Consumers Remain Cautious. Wall Street Journal. Available online: http://www.wsj.com/articles/u-s-consumer-spending-flat-in-april-1433161904 (accessed on 12 September 2016).

56. Meehan, M. Trends for 2016, Part 2: Consumers Find Tech Transformed, Connect With Their Tribes, Reel From Shock. Forbes. 2015. Available online: http://www.forbes.com/sites/marymeehan/2015/12/16/trends-for-2016-part-2-consumers-find-tech-transformed-connect-with-their-tribes-reel-from-shock/ (accessed on 12 September 2016).

57. Walsh, S. Five Trends That Will Change Consumer Behaviour in 2016. The Globe and Mail. 2016. Available online: http://www.theglobeandmail.com/report-on-business/small-business/sb-growth/five-trends-that-will-change-consumer-behaviour-in-2016/article28019355/ (accessed on 12 September 2016).

58. The Global, Socially Conscious Consumer. Available online: http://www.nielsen.com/us/en/insights/news/2012/the-global-socially-conscious-consumer.html (accessed on 12 September 2016).

59. Cautious Consumers Hold Back on Spending in December as Concerns Rise about U.S. Economy. Available online: http://www.latimes.com/business/la-fi-consumer-spending-20160201-story.html (accessed on 23 March 2016).

60. Kreiger, M.; Pearce, J. Environmental life cycle analysis of distributed three-dimensional printing and conventional manufacturing of polymer products. *ACS Sustain. Chem. Eng.* **2013**, *1*, 1511–1519. [CrossRef]

61. Kreiger, M.; Pearce, J. Environmental impacts of distributed manufacturing from 3-D printing of polymer components and products. *MRS Proc.* **2013**, *1492*, 85–90. [CrossRef]

62. Gebler, M.; Uiterkamp, A.; Visser, C. A global sustainability perspective on 3D printing technologies. *Energy Policy* **2014**, *74*, 158–167. [CrossRef]

63. Schwepker, C., Jr.; Cornwell, T. An examination of ecologically concerned consumers and their intention to purchase ecologically packaged products. *J. Public Policy Mark.* **1991**, *10*, 77–101.

64. Minton, A.; Rose, R. The effects of environmental concern on environmentally friendly consumer behavior: An exploratory study. *J. Bus. Res.* **1997**, *40*, 37–48. [CrossRef]

65. Roberts, J. Green Consumers in the 1990s: Profile and Implications for Advertising. *J. Bus. Res.* **1996**, *36*, 217–231. [CrossRef]

66. Ebreo, A.; Hershey, J.; Vining, J. Reducing solid waste linking recycling to environmentally responsible consumerism. *Environ. Behav.* **1999**, *31*, 107–135. [CrossRef]

67. Pedrini, M.; Ferri, L.M. Socio-demographical antecedents of responsible consumerism propensity. *Int. J. Consum. Stud.* **2014**, *38*, 127–138. [CrossRef]

68. Williams, Z. The DIY Consumer Profile: 10 Proven Characteristics. 2016. Available online: https://www. venveo.com/blog/10-characteristics-of-the-diy-consumer (accessed on 12 September 2016).

69. Zaleski, A. 5 Things to Watch for in 3D Printing in 2016. Fortune. 2015. Available online: http://fortune. com/2015/12/31/5-things-to-watch-3d-printing-2016/ (accessed on 12 September 2016).

70. Roberson, D.; Espalin, D.; Wicker, R. 3D Printer Selection: A Decision-Making Evaluation and Ranking Model. *Virtual Phys. Prototyp.* **2013**, *8*, 201–212. [CrossRef]

71. RepRap Printable Part Sources. Available online: http://reprap.org/wiki/Printable_part_sources (accessed on 12 September 2016).

72. Osborn, L.S.; Pearce, J.M.; Haselhuhn, A. A case for weakening patent rights. *John's L Rev.* **2015**, *89*, 1185–1253.

73. Bhavsar, P. Is It Worth Being Green? Green Business Transitions in Northern California. ES196 Senior Research Seminar. University of California: Berkeley, 2011. Available online: http://nature.berkeley.edu/ classes/es196/projects/2011final/Bhavsar_2011.pdf (accessed on 12 September 2016).

74. Hunt, E.; Zhang, C.; Anzalone, N.; Pearce, J. Polymer Recycling Codes for Distributed Manufacturing with 3-D Printers. *Resour. Conserv. Recycl.* **2015**, *97*, 24–30. [CrossRef]

75. Wu, J. Average Credit Card Interest Rates (APR)—2017. Available online: https://www.valuepenguin.com/ average-credit-card-interest-rates (accessed on 30 January 2017).

technologies

MDPI

Review

Overhanging Features and the SLM/DMLS Residual Stresses Problem: Review and Future Research Need

Albert E. Patterson [1,2,*], **Sherri L. Messimer** [1] and **Phillip A. Farrington** [1]

1   Department of Industrial and Systems Engineering and Engineering Management, University of Alabama
     in Huntsville, Technology Hall N143, 301 Sparkman Drive, Huntsville, AL 35899, USA;
     messims@uah.edu (S.L.M.); farrinp@uah.edu (P.A.F.)
2   Department of Industrial and Enterprise Systems Engineering, University of Illinois at Urbana-Champaign,
     Transportation Building 117, 104 South Mathews Avenue, Urbana, IL 61801, USA
*   Correspondence: aep0049@uah.edu or pttrsnv2@illinois.edu; Tel.: +1-256-824-5290

Academic Editors: Salvatore Brischetto, Paolo Maggiore and Carlo Giovanni Ferro
Received: 24 March 2017; Accepted: 10 April 2017; Published: 12 April 2017

Abstract: A useful and increasingly common additive manufacturing (AM) process is the selective laser melting (SLM) or direct metal laser sintering (DMLS) process. SLM/DMLS can produce full-density metal parts from difficult materials, but it tends to suffer from severe residual stresses introduced during processing. This limits the usefulness and applicability of the process, particularly in the fabrication of parts with delicate overhanging and protruding features. The purpose of this study was to examine the current insight and progress made toward understanding and eliminating the problem in overhanging and protruding structures. To accomplish this, a survey of the literature was undertaken, focusing on process modeling (general, heat transfer, stress and distortion and material models), direct process control (input and environmental control, hardware-in-the-loop monitoring, parameter optimization and post-processing), experiment development (methods for evaluation, optical and mechanical process monitoring, imaging and design-of-experiments), support structure optimization and overhang feature design; approximately 143 published works were examined. The major findings of this study were that a small minority of the literature on SLM/DMLS deals explicitly with the overhanging stress problem, but some fundamental work has been done on the problem. Implications, needs and potential future research directions are discussed in-depth in light of the present review.

Keywords: additive manufacturing; 3D printing; metal additive manufacturing; selective laser melting (SLM); direct metal laser sintering (DMLS); metal powder processing

1. Introduction

Additive manufacturing (AM) technologies, commonly known as 3D printing tools, are a family of manufacturing processes that produce solid geometries by "joining [raw] materials to make objects from 3D model data, usually layer upon layer, as opposed to subtractive manufacturing methods" [1]. While most commonly-used and established AM processes use plastics and photopolymers as the initial raw material, a number of AM processes that can process metals (usually in the form of fine powder) are emerging and being rapidly developed and perfected. The availability of such fabrication tools offers great promise to many sectors of manufacturing, especially the aerospace, medical and automotive industries, in their ever-growing quest for lighter, stronger, tougher, more complex and more cost-efficient metal parts.

One of the most promising and flexible of these metal-printing processes is known as selective laser melting (SLM) or direct metal laser sintering (DMLS). The process is known by both names, depending on the geographical area of the user; in the early days of development, "SLM" was most commonly used in Europe and "DMLS" in the USA, but both names have been used synonymously as the technology has matured over the past decade or so. Compared to other metal-melting AM process, such as electron beam melting (EBM), SLM/DMLS is very cost effective, works well with a wide variety of elemental metals and alloys, produces an excellent surface finish, provides excellent feature resolution and is more industrially safe [2–4]. Unfortunately, the SLM/DMLS process is dominated by one serious weakness, which is preventing its more wide-spread acceptance and use as a standard manufacturing process: the tendency of the process to build an unbalanced stress profile into the part between the layers during processing. This has become known as the residual stresses problem and has been the topic of research since the process was first introduced. The collection of both general and regional residual stresses into parts without a way for them to naturally dissipate (as they do in non-metal AM processes) can be a major problem because this can initiate cracks, warpage and delamination if the part is not properly designed or has delicate features, both during and after processing, and can reduce the fatigue strength of the part by a factor of 10 or more when compared to bulk-formed parts [5–8].

These problems are especially apparent and challenging in parts that have overhanging or protruding features, as the stresses tend to build up more seriously in and near these features during printing [5,9]; this can cause severe warping and damage to the features and cause the destruction of the entire part, sometimes before it is even finished printing. Temporary support structure can be used to prevent in-process failure, but using these in SLM/DMLS can come with its own set of problems. With careful part design, use of special support structures for delicate features and various rules-of-thumb developed over the years, the process can be used successfully for specific applications; however, it would be far more useful and trustworthy, more cost efficient and more widely accepted if a general theory of design were available for the parts that will be created using SLM/DMLS.

2. The SLM/DMLS Residual Stresses Problem

According to the U.S. patent for SLM/DMLS, the process is a variation of the powder bed fusion process in which a thin layer of "metallic powder free of binding and fluxing agents" is selectively "heated by [a] laser beam to melting temperature" in order to fuse it into a solid slice of material in the correct shape of the part. The laser beam energy "is chosen in such a way that the layer of metallic powder is fully molten throughout its layer thickness at the point of impact of [the] laser beam" and the laser beam is "guided across a specified area of the powder material layer … in such a way that each run partially overlaps the preceding run" in order to form proper metallic bonds between scans (and between the current layer and previous layers) and therefore produce a homogeneous solid. The entire operation is run in a "protective gas environment" during the described procedures to prevent unwanted reactions and oxidations. Because the powdered material is "free of binding and fluxing agents" and because it is "heated to its melting temperature throughout the layer thickness", the resulting solid has mechanical properties similar to bulk-formed materials [10]. As each layer is selectively melted in this way, the build table in the printer drops down the distance of one layer thickness (20–100 μm), and a wiper deposits a fresh layer of new unmelted powder, starting the whole operation over again. This cycle continues until the part is complete [5–7,11]. Traditionally, only metallic materials could be used with SLM, but some work has been done to extend the process to ceramics and metal/ceramic/polymer composites [12–14]. Figure 1 demonstrates the basic anatomy and process chain for SLM/DMLS.

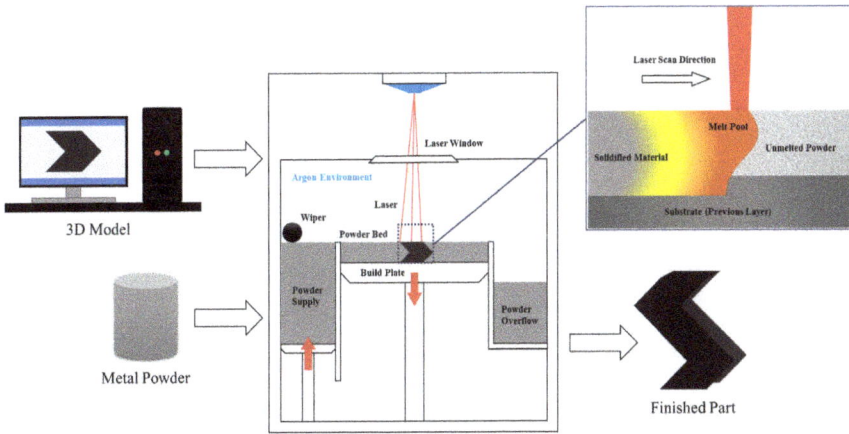

Figure 1. SLM/DMLS process mechanics.

By definition, "residual stresses" are the stresses within a plastically- or elastically-deformed material that remain within the structure after the load that deformed it is removed [15]. In the SLM process, the major source of the residual stresses is the heat cycling as the laser scans across each layer, where previously solidified layers are re-melted and cooled several times at inconsistent levels of heat. When looking at the stress gradients in a particular single layer of the part during heating, the two most important regions are the top of the layer (exposed to the laser) and the interface between the layer and the previous layer (Figure 2).

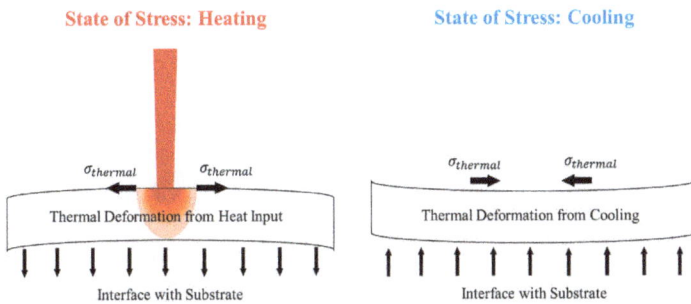

Figure 2. Stress gradients in single layers.

Due to thermal expansion, the top of the layer experiences a tensile stress, while the cooler interface has compressive stresses acting on it. If only one layer were to be printed, this would not be a problem, as the stresses would dissipate naturally once the material cooled. The problem manifests itself when the underlying layers restrict the thermal expansion and contraction of the layers immediately below the melt pool; this can occur several layers deep simultaneously and can happen multiple times to the same layer throughout the build, and the material does not necessarily need to be molten for it to happen. This can cause an elastic compressive strain within the layers, resulting in a stress gradient between the layers [8,9]. Figure 3 demonstrates this graphically. Where the layers are free to move (Figure 3a), the residual stress between the layers is low; it is not zero, however, since some friction will still exist between the layers. Where the layers are restricted from moving by fusion (Figure 3b), the stresses can build up quickly because they are not allowed to move freely and therefore can become warped as the subsequent layers are heated. Figure 4 shows an example finite-element (FEA) model of the thermal deformations during laser scanning; the material shown is six layers (50 μm) of 316 stainless steel, with a 200-W laser input and 24 °C ambient temperature, with the base fixed to the build plate. This model is for concept demonstration only and not a new research tool; no new powder is added in this figure, this is simply the deformation of the material under laser load.

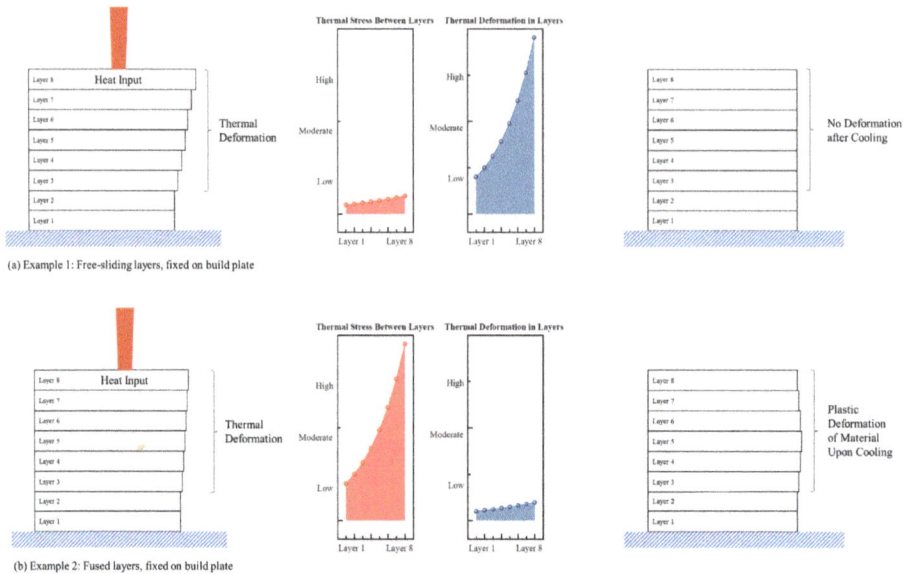

(a) Example 1: Free-sliding layers, fixed on build plate

(b) Example 2: Fused layers, fixed on build plate

Figure 3. Stress between layers.

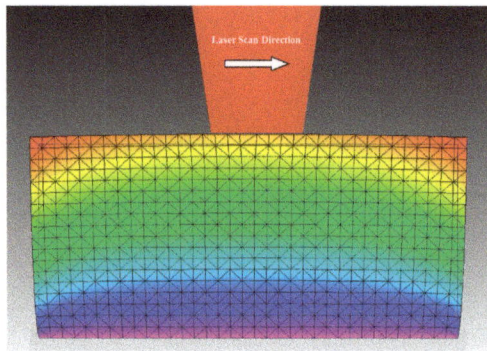

Figure 4. Stress between layers (FEA deformation example).

Several published studies explicitly described the specific mechanics of the stress formulation as described above, most notably in the works of Mercelis and Kruth [9] and Knowles et al. [8]. Other studies that discussed this issue in depth were those by Roberts et al. [16,17], Matsumoto et al. [18], Gu et al. [19], Guo and Leu [4] and Van Belle et al. [20].

There are a number of ways to combat the residual stresses problem when printing very simple parts; most parts created by SLM are physically connected to the build plate at the base, helping to both support and tie down the layers until the part body is large enough to support the stresses.

This is accomplished by fusing the first layer of powder directly to the build plate as if it were the material substrate; this is common knowledge in the world of metal powder manufacturing. Unfortunately, there is little experimentally-based information to be found concerning the effects of the residual stresses on the design of complex parts with overhanging or protruding features (Figure 5). Most of the studies typically discussed in literature searches discuss rule-of-thumb ways to physically prevent the stresses from destroying the parts during printing and are little concerned with trying to understand the mechanics of the stresses and how they directly affect the overhanging features. There was considerable discussion of this problem in the studies by Hussein et al. [21,22], Matsumoto et al. [18], Calingnano [23], Mohanty and Hattel [24], Zeng [25], Li et al. [26] and Gan and Wong [27], but these addressed application-specific issues and did not discuss the problem at the level of feature and part design.

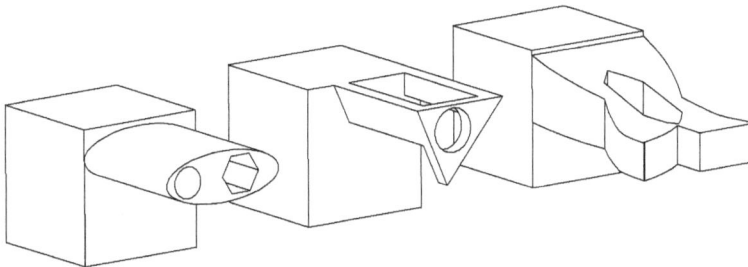

Figure 5. Examples of overhanding and protruding part features.

Some of the opinions commonly heard from practitioners are that the overhanging features are most severely affected by the stresses because they are not physically welded to the build plate during the printing and are thinner and less resistant to thermal shock. Depending on the specific geometry, stress concentrations between the features and the main parts also likely play a role in magnifying the effect. However, there is little rigorous treatment of this in the technical literature to verify if these opinions are indeed true for general cases. The current study hopes to discover more in-depth answers to these questions.

Up to now, the best solution has been to use strong support materials, in spite of some problems using both a powder bed and supports; concerns included the required extra post-processing, extra material use, increased cycle time, increased risk of damage to the part, damage to the finish of the part from support removal and restrictions on the part design to accommodate the support structure; all are issues when using support structures with SLM. Some studies that discuss the pros and cons of support structures to prevent damage in SLM/DMLS well were those performed by Hussein et al. [21,22], Jhabvala et al. [28], Matsumoto et al. [18], Thomas and Bibb [29,30], Wang et al. [31], Kruth et al. [32] and Papadakis et al. [33]. Data from several studies by Hussein et al. [22], Kruth et al. [32], Vora et al. [34] and Patterson et al. [35,36] suggested that the use of rigid support structures during SLM/DMLS for overhanging features may actually cause the residual stresses to be worse than if the overhang had no solid support during printing.

3. Survey of Previous Work

The main goal of this literature review was to identify studies and methods used or in development for SLM/DMLS to combat the negative effects of residual stresses within overhanging and protruding features. To accomplish this, a large number of fundamental sources was collected, sorted into categories and reviewed; they will be discussed in-depth here and in Section 4. This review is not meant to be an annotated bibliography and does not claim to cover every single published work in any particular area. The review simply explores the topic in-depth in order to define the problem and discover the kind of solutions that may be available to deal with it. With this information, future research directions can be identified and guided.

The residual stresses problem has been an obvious problem since the invention of the SLM process and has put a cap on its full and free utilization, so a number of researchers has worked to develop solutions to this problem since the early days of the technology in 1997–2001 [10,18]. Relevant previous work in this area can be categorized into five areas (Figure 6): (1) process modeling and simulation, (2) process control and post-processing methods, (3) experiment development, (4) support structure optimization and (5) design and analysis of overhanging structures. While the great majority of the previous work does not directly address the overhanging structures problem, works that are clearly or potentially relevant to the topic are collected and reviewed in this section, with explicit treatments of the overhang structures addressed at the end of the review.

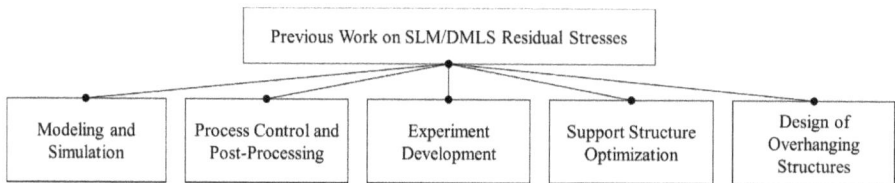

Figure 6. Previous work categories.

3.1. Process Modeling and Simulation

As with any problem solution, a good model is needed for problem understanding before any useful work on the problem can be attempted. A number of models has been developed for the SLM/DMLS process, some general process models and many that model specific aspects of the process. These modeling studies can be sorted into several subcategories, as shown in Figure 7.

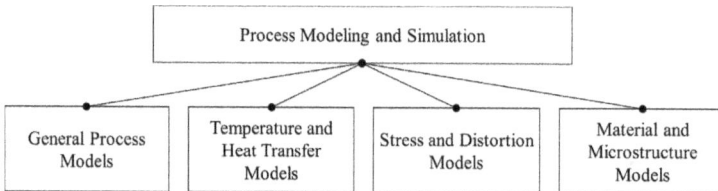

Figure 7. Process modeling and simulation categories.

3.1.1. General SLM/DMLS Process Models

Two of the best-known, trusted and widely-cited general SLM/DMLS process models were produced by Kruth et al. [9,37,38] at the University of Leuven (Belgium) and manufacturing scientists at Lawrence Livermore National Laboratories (LLNL) [39–43] in the United States. Both research teams developed comprehensive analytical thermo-mechanical models of the SLM/DMLS process, which are based on first principles, basic energy balances, phase changes, material properties, material states and part geometries, among many other physical phenomena. Numerous experiments, both physical and numerical, were developed to develop and verify these models, which are considered to be the state-of-the-art in the field. The limiting factor of these models is that they are highly proprietary and not usually available for use by outside research groups and practicing engineers.

Another research group based primarily in Germany, Papadakis et al. [33], proposed a model reduction in order to simplify the creation and running of good finite element models of the thermal and mechanical effects in large parts made by SLM. The study shows that for large parts, the finite element model can be simplified without a significant loss of accuracy and usability. The model considers all of the most important process considerations, such as heat input, molten region geometry, material deposition, phase transformations, heat transfer modes and other effects. The model is well-verified experimentally and is becoming more widely used.

These state-of-the-art models can become very cumbersome to use for more general problems, so Markl and Korner [44] developed a numerical model on multiple length and time scales to model and describe various aspects of the process across numerous applications and parameter sets. Carefully-designed experiments were used to tune and verify the numerical model in real time in order to provide a more comprehensive understanding of the underlying physics of the process. This marriage of simultaneous modeling and experimentation to understand SLM/DMLS provides a larger window and clarifies much of the mystery behind the process for product designers.

3.1.2. Temperature Distribution and Heat Transfer Models

Many excellent heat transfer models of the SLM/DMLS process have been developed; the earliest established models were primarily stress based (see Section 3.1.3), which depended heavily on the heat transfer mechanisms within the material. Early research showed that the heat transfer was an unpredictable quantity in standard stress models, necessitating the development of complex heat transfer models. Many of the models deal with the temperature distribution and gradients within the material during processing using finite element analysis; the model, developed by Contuzzi et al. [45] at the Polytechnic University of Bari in Italy, advanced a simple finite element analysis model to simulate the temperature distribution through the layers during the SLM process; stresses were not directly addressed in the model, but a stress model could easily be derived from the heat transfer model. The model also includes a method for directly modeling the phase change of the materials as the process is being run.

The models produced by Huang et al. [46], Li et al. [47] and Kundakcioglu et al. [48] are similar in nature to [45], but are more theoretically based, make fewer assumptions and heavily consider volume shrinkage and phase changes within the material. Coupled transient heat and mechanical analyses are used in these models. The study by Masoomi et al. [49] combines the theory of several thermal models and gathers significant empirical and experimental data concerning the true heat profile. Roberts et al. [16,17] used a novel finite element analysis method known as "element birth and death" to facilitate modeling the heat gradients and the heat transfer between layers. A numerical experiment was performed, in which the stresses in a single layer were studied in detail, and a very complex FEA model was created of the heat transfer and stresses for a very small area.

Other, more specific, thermal models were developed by Gusarov et al. [50], Li et al. [51], Fu and Guo [52], Shifeng et al. [53] and Heeling et al. [54]. Gusarov et al. modeled the heat transfer in the material, both conductive and radiative, assuming that the laser scan tracks were nonuniform and that the material temperature was unstable. Li et al. varied the scan speed and modeled how this changed the heat profile within the material during processing, while Fu and Guo modeled the thermal history in the material as a function of layer buildup, which varied significantly with time. The mechanics of the melt pool, its boundaries. and its influence on the surrounding material was modeled by Shifeng et al. and Heeling et al. using finite element methods.

3.1.3. Stress and Distortion Models

The primary purpose of much SLM/DLMS research is the accurate and effective modeling of part distortion and deformation during processing in order to produce good quality finished parts. The earliest examples of a stress model for SLM/DMLS were developed by Matsumoto et al. [18,55] at Osaka University in Japan and first published in 2001–2002. Kruth et al. at University of Leuven in Belgium have also worked extensively on this problem [9,32,37–39,56–58] and over time developed one of the most well-respected general SLM/DMLS models in the world, as discussed previously. Other important stress and distortion models that have been developed can be classified into two categories:

models of single layer processing and models of bulk (multiple layer) processing. Single layer models analyze the stress effects in just one layer of the part, while bulk models treat several layers or even an entire part at once. In general, the single layer models are more detailed, but the bulk models give a more system-level view of the processing effects.

The most widely-cited single layer stress models using finite element analysis are those developed by Hussein et al. [21], Matsumoto et al. [18], Contuzzi et al. [45] and Dai and Gu [59]. Wu et al. [60] proposed a model that analyzed the stresses within a single layer of powder as it solidifies, unlike the others, which were based on stresses within the solid materials. A variety of bulk (multiple layer) models exist and can be divided into four groups: first principles and analytical models, computational FEA studies that are verified using simple beam deformation experiments, finite element models built in commercial software (such as ANSYS) and multiscale modeling to predict part distortion. First principles models, both simple and complex, were proposed by Patterson et al. [35,36] and Fergani et al. [61], all of which were demonstrated and verified using various numerical experiments and comparisons to published experimental data. Examples of computational studies that were verified using various simple part deformation experiments were those performed by Vrancken et al. [62], Zinovieva et al. [63], Liu et al. [64] and Safronov et al. [65]. Stress models built using ANSYS include those models developed by Zaeh and Branner [66] and Gu and He [67]. The studies by Li et al. [68,69], Parry et al. [70] and Vastola et al. [71] were multiscale finite element models for fast and efficient prediction of part distortion, primarily intended to inform part designers and engineers.

3.1.4. Material and Microstructure Models

The presence of residual stresses within the material clearly influences the way that the material solidifies and forms the microstructure during cooling. Several studies have explored this in depth from several different perspectives, including microstructure evolution, the effects of some specific process parameters on the microstructure, evaluation of bonding issues related to surface roughness and modeling small defects in the material structure during processing.

Some studies that examined microstructure evolution were those by Liu et al. [72], Toda-Caraballo et al. [73], Thijs et al. [38], Chen et al. [74], Mertins et al. [56,75] and Vastola et al. [76]. Liu et al. and Thijs et al. examined the residual stress evolution at the microstructure scale using Vickers hardness tests and concluded that the residual stresses within the microstructure were greatest in the overlapping regions between scan tracks, but were heavily dependent on scan speed and heat profile. Toda-Caraballo et al. examined the influence of the residual stresses in the material on the recrystallization behavior in new solid material as the part was built. Chen et al. examined and modeled the basic thermal behavior of the material during processing at the microstructure level. An examination of an out-of-equilibrium microstructure was examined by Mertins et al., finding that defects in the material were produced both by poor melting/cooling dynamics and a lack of complete melting of some powder during processing. Vastola et al. produced a model of microstructural evolution in both SLM/DMLS and the electron beam melting process and captured some of the process-specific characteristics of the two to explain experimentally-observed differences in microstructure.

Alyoshin et al. [77] examined the microstructural problems when using SLM/DMLS to process materials with poor weldability and developed a method for finding and modeling microcracks in the material. Alloys with poor weldability typically have a low fatigue life, as the recrystallization of the material grains is poor. The researchers were able to increase the fatigue life, particularly in the plastic region, by relaxing the residual stresses using an argon-based treatment to better form the grains during processing.

3.2. Process Control and Post-Processing

Direct process control and post-processing are the most common and preferred methods of dealing with the residual stresses in practice. Several categories of solutions (Figure 8) have been developed, including process input control, environment control, in situ monitoring and feedback control, process parameter optimization and post processing.

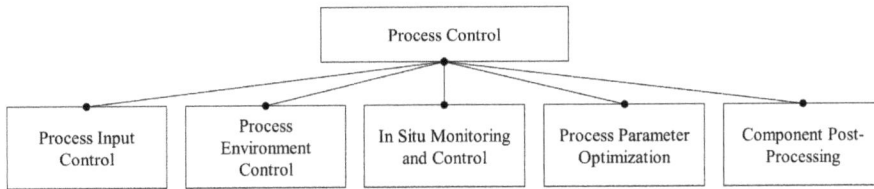

Figure 8. Process control and post-processing categories.

3.2.1. Process Input Control

The basic goal of process input control is to parameterize and control the values of the input parameters, such as laser power, scan speed and other factors, in order to obtain the best possible processing results. While this is the most common technique, besides post-processing, to optimize parts, it depends mostly on the experience and intuition of the user and is usually not applicable to general problems using SLM/DMLS. While many case studies and part- and machine-specific solutions have been published, the best documented and most widely-cited solutions that analyze residual stresses were those by Kruth et al. [32], Carter et al. [78], Zhang et al. [79], Abe et al. [55], Bo et al. [80], Shiomi et al. [81], Yasa and Kruth [82] and Mumtaz and Hopkinson [83].

Kruth et al. and Carter et al. explored the effects on the thermal stresses of modifying the length and orientation of the laser scan vectors, performing pre- and post-scanning and island scanning, varying the layer thickness, heating the base plate and heat treating the final parts. The results showed that all of these modifications to the process produced improved thermal stress values, particularly modification of the scan parameters and reducing the temperature gradient by preheating the build plate. Zhang et al. looked at the effects of the laser parameters, powder setup, environmental conditions and preheating on the quality of the final parts. Shiomi et al. explored the influence of three major factors: heat treating the part after printing (improvement of 70%), heating the powder bed during printing (40% improvement) and re-scanning each layer before printing the next one (55% improvement). Abe et al. and Bo et al. suggested that the scan pattern of the laser can be designed so that the residual stresses can be "designed" and contoured to dissipate naturally or even provide material advantages for the part. Yasa and Kruth analyzed the value of scanning each layer more than once (re-melting) and found that this additional operation significantly reduced the residual stresses by "massaging" them out of the material. Mumtaz and Hopkinson found that using a pulsed laser in SLM resulted in better control over the structure and features, as the power output of the laser was easier to control.

Other useful studies that varied the processing parameters to control residual stresses include those by Tolosa et al. [84], Brandl et al. [85], Edwards and Ramulu [86], Guan et al. [87] Yadroitsev and Smurov [88], Yadroitsev et al. [89], Cheng et al. [90], Xia et al. [91], Lu et al. [92] and Yu et al. [93]. Tolosa et al., Brandl et al. and Edwards and Ramulu varied the orientation of build samples to study the effects of the material anisotropy on the mechanical properties of parts. Build orientation was also studied by Guan et al., as well as various layer thicknesses, overlap rates and hatch angles. Yadroitsev and Smurov studied the influence of surface roughness on bond strength between layers. Yadroitsev et al. studies the combination of pre-heating the build plate and varying the scan speed. Various adjustments to the laser settings were studied by Cheng et al., Xia et al. and Lu et al. Yu et al. examined the influence of various processing parameters on laser penetration depth and melting/re-melting densification during selective laser melting of difficult aluminum alloys.

3.2.2. Process Environment Control

In addition to parametrizing the basic input parameters for the process, modifying the chamber environment seems to have a positive effect on the residual stresses. These controls primarily consisted of chamber temperature control, using inert gases to prevent oxidation and reduce temperature gradients in the powder bed. Jia and Gu [94], Dai and Gu [59] and Dadbakhsh et al. [95] looked at the effect of having oxygen in the environment during printing and ways to eliminate it. Dai and Gu and Dadbakhsh et al. suggested running an inert gas through the powder bed during the process to prevent oxidation between the layers of the part and produce a more uniform temperature throughout. Ladewig et al. [96] examined the use of the inert gas to deal with metal splatter and to flush out process by-products and trash. Buchbinder et al. [97] and Mertens et al. [75] examined the ways to effectively pre-heat the powder and build plate to reduce the likelihood of stresses.

3.2.3. In-Situ Monitoring and Control

SLM/DMLS is a notoriously difficult process to monitor and control during processing due to its complex nature and the need for a perfectly clean and oxygen-free environment to function properly. Methods for monitoring and controlling the process are clearly valuable and will increase the usefulness and breadth of experimentation with the process in the future. Two major systems for real-time process control have been proposed and are in development by Craeghs et al. [98–101] and Devesse et al. [102]. Both systems use a system of optical sensors to collect information about the progress of the part build and to send temperature data to a processor that can control and make modifications to the process parameters in real time. Both of these systems can help to control the process in real time and adjust the parameters as needed; general monitoring and testing technologies are discussed later in the section on experimental development.

3.2.4. Process Parameter Optimization

Most of the previous research on the influence of process parameters on the stresses and deformations in SLM/DMLS have been parametric studies, where effects from adjusting parameters were measured. A different type of parameter study that has been published is the optimization of parameters to gain the best possible solution before the processing begins. The major works in this area have been from Pacurar et al. [103], Casalino et al. [104] and Aboutaleb et al. [105]. Pacurar et al. developed a system for automatically generating process parameters based on models of the process, while Casalino et al. use a statistical optimization technique, and Aboutaleb et al. uses a knowledge database approach, which catalogs the results from previous studies and selects the best parameters based on these results.

3.2.5. Part Post-Processing

The most common way to deal with the residual stresses within SLM/DMLS parts is to post-process them after building. This solution is very simple, as it makes use of existing technologies and does not require special knowledge or modification to the SLM/DMLS process itself. However, post-processing can add to the time required to produce the parts and dramatically increase the cost, while the post-processing itself may not fully remove the stresses and may expose them, destroying the part in the process. The normal forms of post-processing for SLM/DMLS are heat treatment and hot isostatic pressing (HIP) [106–116], but methods such as shot-peening have been successfully used, as well [117].

3.3. Experiment Development

Experimental methods that can be applied to SLM/DMLS are very valuable, as the process is very difficult to monitor and control using traditional methods. Methods that have been developed or adapted for use with SLM/DMLS can be categorized as shown in Figure 9.

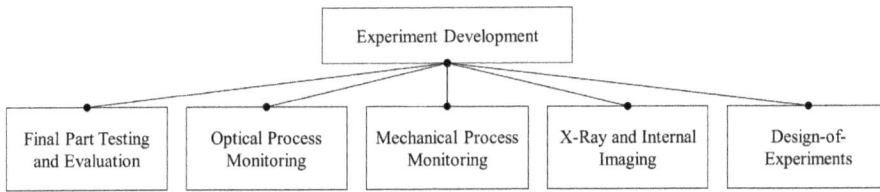

Figure 9. Experiment development categories

3.3.1. Final Parts Testing and Evaluation

Final part testing and evaluation is very important, as any SLM/DMLS parts that are used commercially, or in government, or military use must be tested and certified for duty. A testing standard should be developed for this, but there is yet to be one available. Previous work has been done in developing part evaluation techniques, including ultrasonic testing [118] and various methods for tracking crack growth in the parts after processing [5,6].

3.3.2. Optical Process Monitoring

Since SLM/DMLS is so difficult to monitor and control using traditional methods that new methods are very useful. Optical methods are the easiest to use during experimentation, as they are usually non-disruptive to the process and can be applied externally without modifying the process or equipment. A number of excellent optimal monitoring systems has been developed, in particular those by Craeghs et al. [98–101], Kleszczynski et al. [119], Clijsters et al. [120], Chivel [121], Grasso et al. [122], Hirsh et al. [123], Kanko et al. [124] and Lott et al. [125]. Infrared thermography systems for SLM/DMSL have been developed by Rodriguez et al. [126] and Smurov et al. [127].

3.3.3. Mechanical Process Monitoring

While optical process monitoring is less disruptive, the major disadvantages are in calibrating the imaging devices and in monitoring non-surface phenomena. Some methods have been developed to deal with this problem by effectively using strain gauges on or near the parts to monitor the materials during processing, such as those by Knowles et al. [8] and Casavola et al. [128]. Others fix the parts to larger bodies, which contain various sensors in order measure deformation in real time, represented by the methods developed by Yadroitsava and Yadroitsev [129], Dunbar et al. [130] and Havermann et al. [131].

3.3.4. X-ray and Internal Imaging

Yadroitsava et al. [132] developed an X-ray diffraction technique to study residual stresses within parts during and after processing, overcoming some of the challenges of the mechanical monitoring methods while providing the hands-off benefits of thermal monitoring.

3.3.5. Design-of-Experiments

To more effectively study process parameter effects on residual stresses and deformations during processing, designed experiments should be used. These are carefully formulated experimental approaches and tools that allow valid statistical analysis of data collected from experiments and reduce the number of experiments needed to draw defendable conclusions from processing data. While still under development, methods by Patterson et al. [35,36] and Protasov et al. [133] appear promising for future experimental design in SLM/DMLS.

3.4. Support Structure Optimization

Structural supports are typically needed in order to prevent the failure of unsupported overhanging features, as well as many other complex types of features. This, however, is a necessary nuisance that must be tolerated to utilize the design freedom of this AM process; the extra time required to cut, grind or mill off the support structures, the extra material used (which is wasted), the longer print time, the damage to the surface finish when the structure is removed, the extra time required to design the part to accommodate the structure and the design of the structure itself are some of the irritations that come with using SLM to create parts with overhanging features. Thankfully, work has been done to simplify the job and reduce the impact of the support structures, while reaping the full benefit of using the structures. Many studies in this area exist, but only the fundamental papers that present new and novel methods (as opposed to case studies) are shown here. It should be noted that optimization of removable support materials and the design of overhanging features are separate topics of study and therefore will be discussed separately in a later section of the paper.

Some of the most fundamental work in this area was done by Sundar et al. [134], Jhabvala et al. [28], Hussein et al. [21,22], Maliaris et al. [135] and Strano et al. [136]. Sundar et al. found that printing the part on top of a wire mesh made removal from the build plate easier, facilitated the creation of delicate features and thin walls, reduced the time needed to cut off the support structure and created a buffer to prevent damage to the part itself from the removal. Jhabvala et al. built support structures using a pulsed laser, which has a number of advantages, including support material that is not full density and is soft compared to the rest of the body, but is strong enough to handle the stresses and heat transfer. This creates a structure that provides support, but is very easy to remove during post-processing. The laser itself was set to both full-power and pulse modes as needed, the full-power mode creating the part and the pulsed setting creating the support structure. Hussein et al., Maliaris et al. and Strano et al. experimented with using delicate cellular lattice structures as supports; the advantages to this are material savings, easier removal from the part, and some time savings compared to methods using solid support structures. However, this takes extra time to design.

3.5. Overhanging Feature Design

When collecting sources for the other sections, several references to and discussions about designing overhanging structures in SLM were found. While the number of sources directly related to the overhang problem is far smaller than many of the other SLM/DMLS topics, progress is being made to address the problem. Additional searches were also performed in several journal databases and academic search engines, and this uncovered several relevant papers, in addition to studies that have already been discussed.

In work discussed earlier, four studies discussing various aspects of overhanging structures were Calignano [23], Mohanty and Hattel [24] and Patterson et al. [35,36]. Calignano suggests avoiding overhanging structures in design as much as possible; however, when they are unavoidable, special support structures should be designed to have the minimum possible contact with the overhang. A detailed discussion about overhang design is presented, as well, with several case studies showing great improvement in warping when following the design rules. Mohanty and Hattel looked at the influence of scan orientation on the quality of the overhang structures, with and without support, and conducted a detailed error and sensitivity analysis. Patterson et al. suggested developing a factorial-based design-of-experiments (DOE) approach to stresses and deformation in 90-degree overhangs, both supported and unsupported. The studies by Patterson et al. were unique because they considered the influence of geometric stress concentrations, as well as the normal part deformation under thermal load. A detailed numerical study and comparison to published experimental data showed that the stress concentration had a very large influence, at least as much as the laser power, on the stress and deformation. The DOE approach also allowed the calculation of parameter effects and interactions, allowing a multi-dimensional analysis of the problem. The simple thermal model used in these studies needs more development, refinement and verification, but the results gathered match expected results from other studies.

Other works that addressed the concern of overhanging feature stresses in SLM/DMLS were those by Wang et al. [31,137,138], Cloots et al. [139], Fox et al. [140] and Kruth et al. [141]. The focus of the studies performed by Wang et al. was the design of curved overhanging parts and parts set at small angles, so designed that they did not need significant support material. The primary analyses were done to determine the settings and design to accomplish the best possible surface finish for the parts. Cloots et al. proposed a method like that of Calignano [23], except that structures were lattice networks instead of support points. The study also focused on the number of layers needed to provide a stable part overhang with the goal of minimizing the need for the supports altogether. A small case study was also done to show that the overhang design technique used by Cloots et al. could be used to stack parts and provide more dense part packing on the build plate. Similar to the studies by Wang et al., Fox et al. was interested in the surface finish of part overhangs and studied empirical relationships between process parameters, overhang angle and surface roughness. The study by Kruth et al. was a benchmark study where a number of different geometries were built, including overhanging features, and the results were compared between SLM/DMLS and other processes.

4. Discussion and Future Research Need

In this study, a deep and detailed literature review was done to collect previous works related to the effect of SLM/DMLS residual stresses on delicate overhanging and protruding features. Unfortunately, little work has been done to explicitly address this issue or to even understand and model it properly; most of the conclusions about overhanging featured were limited to numerical studies and part-specific design case studies. Clearly, much research effort is needed in this area in the future. In the process of examining the literature for works related to stresses in overhanging features, a large body of work related to the general residual stresses problem was collected. The work reviewed was divided into a set of 16 categories:

1. General SLM/DMLS process models
2. Heat transfer models
3. Stress and distortion models
4. Material and microstructure models
5. Direct process input control
6. Direct environment control
7. Hardware-in-the-loop monitoring
8. Process parameter optimization
9. Part post-processing
10. Part evaluation method development
11. Optical process monitoring
12. Mechanical process monitoring
13. Internal imaging method development
14. Design-of-experiments
15. Support structure optimization
16. Overhang feature design

Each of the first 15 categories has existing tools that can be applied to further work in the area of overhanging feature design in the future, such as modeling, process control, post-processing, part evaluation methods, design-of-experiments and support structure optimization. However, none of these tool sets are complete in themselves and require additional refinement and development in the future to become more powerful, useful and reliable. This review was very helpful in uncovering some of the major gaps and needs for future research in this area. Some suggestions on future directions and projects are:

(1) Process models clearly are useful in analyzing overhanging and other complex structures; however, great care must be taken to make sure they accurately model the material conditions in the presence of overhanging structures. Some aspects need further consideration in future research when used for overhanging and other complex features, particularly in the mechanical and heat responses of the overhanging features. These features may act like mechanical springs, deforming in a non-linear fashion, and could introduce extra vibrations into the material during processing and use. The overhanging features will also be subjected to different heat conditions

than the rest of the part; the features will generally be thinner and subjected to much faster energy transfer from the laser (and therefore, much more severe stresses).

(2)   Something that was not encountered in any detail in the reviewed literature is the presence of regions of stress concentration in and near overhanging features. This, combined with unknown heat effects, puts into question the results from existing models with complex geometry, questions that should be analyzed and answered.

(3)   Most of the previous work in verifying the models was the completion of numerical and parametric studies; formally-designed experiments should be used to further verify these models, as they are capable of analyzing both the main effects from the input factors and the interactions between these factors. While they are more expensive than parametric studies and require detailed planning before research begins, the use of interaction analysis will aide in the quick identification and tracking of error factors in the models. This will allow a higher confidence over the needed analysis range and therefore more trustworthy models.

(4)   Another major concern in using models for this manufacturing process is that the best and most trusted models for SLM/DMLS are proprietary or government lab-owned and not available for use and improvement by the SLM/DMLS community. This can stunt the growth of accurate general-use design models, which will be essential when developing formal design-for-manufacturability methods. Greater access and transparency with these models should be pursued in the future. At the least, those who own and develop the proprietary models should publish technical works guiding the formation of more public-use models.

(5)   To simplify the design process, a method should be developed to identify the "dominating" factors within the SLM/DMLS build plan for particular designs. Using this, the part can be redesigned or the decision can be made by the designer that some or all of the "dominated" factors can be safely ignored (as is often done in engineering optimization problems [142]). This will create a much more efficient system, but care should be taken with this task to make sure that the ignored factors are indeed dominated and not just weak factors in the application range.

(6)   Alongside developing post-processing techniques, direct control of the process parameters is the usual first line of defense when dealing with residual stresses in SLM/DMLS, particularly in complex and overhanging part features. The ability to control the process parameters simplifies the processing of the complex geometries and allows custom, optimal parameters for particular applications. There are still limitations in this, however, which need to be addressed: In most cases, the custom process parameters are set by the user before the processing begins. In situ monitoring and hardware-in-the-loop (HWIL) systems partially solve this problem, but still rely on the detection of some anomaly or defect in the part before process parameters are modified. Even if the form of the part can be saved, it is typically scrap and not trustworthy for its original purpose. Some sort of an anticipatory system is needed, perhaps based on a combination or the digital build path progress and preliminary scanning of the powder layer for potential defects. While this could make the process much slower, it could dramatically reduce the failure rate; the slower build speed may also assist in the creation of overhanging features by reducing the magnitude of the thermal shock experienced by the feature during scanning.

(7)   An in situ system for monitoring the quality of the fresh powder layer itself (prior to scanning each layer) could be an important advancement and could use existing technology. The process would need to be stopped for a scan between each layer, which could be a simple roughness measurement with a laser or could be an ultrasound or X-ray scan. The ultrasound scan might require disturbing the powder bed somewhat, but the settling effect could prevent air pockets and help the layers be more uniform in thickness. The powder bed would be more tightly packed, as well, reducing (but not eliminating) the need for support material for some overhang geometries.

(8)   A system could also be developed that controls laser power as a function of the material thickness at a particular scan location. An optimal minimum material thickness could be determined

experimentally as a function of laser power. When the laser encounters thin sections of the geometry, the power will be reduced to avoid thermal shock to the material and provide a consistent amount of heat flux into the material.

Finally, in order for the part made using SLM/DMLS to be useful in the real world, there must be a system for testing, verifying and certifying the parts. If no other major research needs described in this paper are attempted, the formation of technical testing and quality standards should be a priority for the SLM/DMLS community. Of all of the potential projects described here, the development of these standards is the most urgent and critical; even initial and draft guidelines based on current knowledge are a starting place from which excellent documents can be developed.

Acknowledgments: No external funding or grants were provided for the completion of this study.

Author Contributions: Albert E. Patterson collected and summarized articles, created the figures and completed the author metrics study. Sherri L. Messimer and Phillip A. Farrington guided the direction of the study, provided additional articles and carefully checked the study for errors. Albert E. Patterson wrote the basic manuscript, which was edited by all three authors.

Conflicts of Interest: The authors declare no conflict of interest.

1. ASTM F2792-12a. *Standard Terminology for Additive Manufacturing Technologies (Withdrawn 2015; Enforceable Until 2020 Per ASTM Guidelines)*; ASTM International: West Conshohocken, PA, USA, 2012. [CrossRef]
2. Gibson, I.; Rosen, D.W.; Stucker, B. *Additive Manufacturing Technologies: Rapid Prototyping to Direct Digital Manufacturing*, 1st ed.; Springer: New York, NY, USA, 2010; pp. 103–135.
3. Gebhardt, A. *Understanding Additive Manufacturing: Rapid Prototyping, Rapid Manufacturing, and Rapid Tooling*; Hanser: Cincinnati, OH, USA, 2012; pp. 40–44.
4. Guo, N.; Leu, M.C. Additive Manufacturing: Technology, Applications, and Research Needs. *Front. Mech. Eng.* **2013**, *8*, 215–243. [CrossRef]
5. Leuders, S.; Thöne, M.; Riemer, A.; Niendorf, T.; Tröster, T.; Richard, H.A.; Maier, H.J. On the mechanical behavior of titanium alloy TiAl6V4 manufactured by selective laser melting: Fatigue resistance and crack growth performance. *Int. J. Fatigue* **2013**, *48*, 300–307. [CrossRef]
6. Riemer, A.; Leuders, S.; Thöne, M.; Richard, H.A.; Tröster, T.; Niendorf, T. On the fatigue crack growth behavior in 316L stainless steel manufactured by selective laser melting. *Eng. Fract. Mech.* **2014**, *120*, 15–25. [CrossRef]
7. Rafi, H.K.; Starr, T.L.; Stucker, B.E. A comparison of the tensile, fatigue, and fracture behavior of Ti-6Al-4V and 15–5 PH stainless steel parts made by selective laser melting. *Int. J. Adv. Manuf. Technol.* **2013**, *69*, 1299–1309. [CrossRef]
8. Knowles, C.R.; Becker, T.H.; Tait, R.B. Residual Stress Measurements and Structural Integrity Implications for Selective Laser Melted TI-6AL-4V. *South Afr. J. Ind. Eng.* **2012**, *23*, 119–129. [CrossRef]
9. Mercelis, P.; Kruth, J.-P. Residual Stresses in Selective Laser Sintering and Selective Laser Melting. *Rapid Prototyp. J.* **2006**, *12*, 254–265. [CrossRef]
10. Meiners, W.; Wissenbach, K.; Gasser, A. Selective laser sintering at melting temperature. U.S. Patent No. 6,215,093, 10 April 2001.
11. Dutta, B.; Froes, F.H. Additive Manufacturing of Titanium Alloys. *Adv. Mater. Process.* **2014**, *172*, 18–23. Available online: http://amp.digitaledition.asminternational.org/i/250000-feb-2014/20 (accessed on 11 April 2017).
12. Wilkes, J.; Hagedorn, Y.-C.; Meiners, W.; Wissenbach, K. Additive Manufacturing of ZrO_2-Al_2O_3 Ceramic Components by Selective Laser Melting. *Rapid Prototyp. J.* **2013**, *19*, 51–57. [CrossRef]
13. Hao, L.; Dadbakhsh, S.; Seaman, O.; Felstead, M. Selective Laser Melting of a Stainless Steel and Hydroxyapatite Composite for Load-Bearing Implant Development. *J. Mater. Process. Technol.* **2009**, *209*, 5793–5801. [CrossRef]
14. Shishkovsky, I.; Yadroitsev, I.; Bertrand, P.; Smurov, I. Alumina–Zirconium Ceramics Synthesis by Selective Laser Sintering/Melting. *Appl. Surf. Sci.* **2007**, *254*, 966–970. [CrossRef]

15. Black, J.T.; Kohser, R.A. *DeGarmo's Materials and Processes in Manufacturing*, 10th ed.; Wiley: Hoboken, NJ, USA, 2008; pp. 89–96.
16. Roberts, I.A.; Wang, C.J.; Esterlein, R.; Stanford, M.; Mynors, D.J. A Three-Dimensional Finite Element Analysis of the Temperature Field During Laser Melting of Metal Powders in Additive Layer Manufacturing. *Int. J. Mach. Tools Manuf.* **2009**, *49*, 916–923. [CrossRef]
17. Roberts, I.A. Investigation of Residual Stresses in the Laser Melting of Metal Powders in Additive Layer Manufacturing. Doctoral Dissertation, University of Wolverhampton, Wolverhampton, UK, 2009. Available online: http://wlv.openrepository.com/wlv/handle/2436/254913 (accessed on 11 April 2017).
18. Matsumoto, M.; Shiomi, M.; Osakada, K.; Abe, F. Finite Element Analysis of Single Layer Forming on Metallic Powder Bed in Rapid Prototyping by Selective Laser Processing. *Int. J. Mach. Tools Manuf.* **2002**, *42*, 61–67. [CrossRef]
19. Gu, D.D.; Meiners, W.; Wissenbach, K.; Poprawe, R. Laser Additive Manufacturing of Metallic Components: Materials, Processes and Mechanisms. *Int. Mater. Rev.* **2012**, *57*, 133–164. [CrossRef]
20. Van Belle, L.; Vansteenkiste, G.; Boyer, J.-C. Investigation of Residual Stresses Induced During the Selective Laser Melting Process. *Key Eng. Mater.* **2013**, *554*, 1828–1834. [CrossRef]
21. Hussein, A.; Hao, L.; Yan, C.; Everson, R. Finite Element Simulation of the Temperature and Stress Fields in Single Layers Built Without-Support in Selective Laser Melting. *Mater. Des.* **2013**, *52*, 638–647. [CrossRef]
22. Hussein, A.; Hao, L.; Yan, C.; Everson, R.; Young, P. Advanced Lattice Support Structures for Metal Additive Manufacturing. *J. Mater. Process. Technol.* **2013**, *213*, 1019–1026. [CrossRef]
23. Calignano, F. Design optimization of supports for overhanging structures in aluminum and titanium alloys by selective laser melting. *Mater. Des.* **2014**, *64*, 203–213. [CrossRef]
24. Mohanty, S.; Hattel, J.H. Improving accuracy of overhanging structures for selective laser melting through reliability characterization of single track formation on thick powder beds. *Proc. SPIE* **2016**, *9738*. [CrossRef]
25. Zeng, K. Optimization of Support Structures for Selective Laser Melting. Doctoral Dissertation, University of Louisville, Louisville, KY, USA, 2015. [CrossRef]
26. Li, Z.; Zhang, D.Z.; Dong, P.; Kucukkoc, I. A lightweight and support-free design method for selective laser melting. *Int. J. Adv. Manuf. Technol.* **2016**, 1–11. [CrossRef]
27. Gan, M.X.; Wong, C.H. Practical support structures for selective laser melting. *J. Mater. Process. Technol.* **2016**, *238*, 474–484. [CrossRef]
28. Jhabvala, J.; Boillat, E.; André, C.; Glardon, R. An Innovative Method to Build Support Structures with a Pulsed Laser in the Selective Laser Melting Process. *Int. J. Adv. Manuf. Technol.* **2012**, *59*, 137–142. [CrossRef]
29. Thomas, D.; Bibb, R. Identifying the Geometric Constraints and Process Specific Challenges of Selective Laser Melting. In Proceedings of the Time Compression Technologies Rapid Manufacturing Conference, Coventry, UK, October 2008; [CD-ROM]; Rapid News Publications: Coventry, UK, 2008.
30. Thomas, D.; Bibb, R. Baseline build-style development of Selective Laser Melting high density functional parts. In *8th National Conference on Rapid Design, Prototyping & Manufacture*; CRDM/Lancaster University: High Wycombe, UK, 2007; pp. 105–114, ISBN: 9-780948314-537.
31. Wang, D.; Yang, Y.; Zhang, M.; Lu, J.; Liu, R.; Xiao, D. Study on SLM Fabrication of Precision Metal Parts with Overhanging Structures. In Proceedings of the 2013 IEEE International Symposium on Assembly and Manufacturing (ISAM), Xi'an, China, 30 July–2 August 2013; pp. 222–225.
32. Kruth, J.-P.; Deckers, J.; Yasa, E.; Wauthlé, R. Assessing and Comparing Influencing Factors of Residual Stresses in Selective Laser Melting Using a Novel Analysis Method. *Proc. Inst. Mech. Eng.* **2012**, *226*, 980–991. [CrossRef]
33. Papadakis, L.; Loizou, A.; Risse, J.; Bremen, S.; Schrage, J. A Computational Reduction Model for Appraising Structural Effects in Selective Laser Melting Manufacturing. *Virtual Phys. Prototyp.* **2014**, *9*, 17–25. [CrossRef]
34. Vora, P.; Mumtaz, K.; Todd, I.; Hopkinson, N. AlSi12 in-situ alloy formation and residual stress reduction using anchorless selective laser melting. *Addit. Manuf.* **2015**, *7*, 12–19. [CrossRef]
35. Patterson, A.E. Design of Experiment to Analyze Effect of Input Parameters on Thermal Stress and Deformation in Overhanging Part Features Created with the SLM Additive Manufacturing Process. Master of Science Thesis, University of Alabama, Huntsville, AL, USA, 2014. Available online: http://gradworks.umi.com/15/89/1589147.html (accessed on 11 April 2017).

36. Patterson, A.E.; Messimer, S.L.; Farrington, P.A.; Carmen, C.L.; Kendrick, J.T. Understanding Overhang Feature Processing in Selective Laser Melting: Experiment Model Construction. *Int. J. Prod. Manag. Eng.* **2017**, in press.

37. Kruth, J.-P.; Mercelis, P.; van Vaerenbergh, J.; Froyen, L.; Rombouts, M. Binding mechanisms in selective laser sintering and selective laser melting. *Rapid Prototyp. J.* **2005**, *11*, 26–36. [CrossRef]

38. Thijs, L.; Verhaeghe, F.; Craeghs, T.; van Humbeeckm, J.; Kruth, J.-P. A study of the microstructural evolution during selective laser melting of Ti–6Al–4V. *Acta Materialia* **2010**, *58*, 3303–3312. [CrossRef]

39. Lawrence Livermore National Laboratory, Metal Additive Manufacturing. Available online: https://manufacturing.llnl.gov/additive-manufacturing/metal-additive-manufacturing (accessed on 22 March 2017).

40. Khairallah, S.A.; Anderson, A.T. Mesoscopic Simulation Model of Selective Laser Melting of Stainless Steel Powder. Available online: https://e-reports-ext.llnl.gov/pdf/769379.pdf (accessed on 19 March 2017).

41. Khairallah, S.A.; Anderson, A.T. Mesoscopic Simulation Model of Selective Laser Melting of Stainless Steel Powder. *J. Mate. Process. Technol.* **2014**, *214*, 2627–2636. [CrossRef]

42. Hodge, N.E.; Ferencz, R.M.; Solberg, J.M. Implementation of a thermomechanical model for the simulation of selective laser melting. *Comput. Mech.* **2014**, *54*, 33–51. [CrossRef]

43. Hodge, N.E.; Ferencz, R.M.; Vignes, R.M. Experimental comparison of residual stresses for a thermomechanical model for the simulation of selective laser melting. *Addit. Manuf.* **2016**, *12*, 159–168. [CrossRef]

44. Markl, M.; Korner, C. Multiscale Modeling of Powder Bed-Based Additive Manufacturing. *Annu. Rev. Mater. Res.* **2016**, *46*, 93–123. [CrossRef]

45. Contuzzi, N.; Campanelli, S.L.; Ludovico, A.D. 3D Finite Element Analysis in the Selective Laser Melting Process. *Int. J. Simul. Model* **2011**, *10*, 113–121. [CrossRef]

46. Huang, Y.; Yang, L.J.; Du, X.Z.; Yang, Y.P. Finite element analysis of thermal behavior of metal powder during selective laser melting. *Int. J. Therm. Sci.* **2016**, *104*, 146–157. [CrossRef]

47. Li, Y.; Zhou, K.; Tor, S.B.; Chua, C.K.; Leong, K.F. Heat transfer and phase transition in the selective laser melting process. *Int. J. Heat Mass Transf.* **2017**, *108*, 2408–2416. [CrossRef]

48. Kundakcioglu, E.; Lazoglu, I.; Rawal, S. Transient thermal modeling of laser-based additive manufacturing for 3D freeform structures. *Int. J. Adv. Manuf. Technol.* **2016**, *85*, 493–501. [CrossRef]

49. Masoomi, M.; Gao, X.; Thompson, S.M.; Shamsaei, N.; Bian, L.; Elwany, A. Modeling, Simulation and Experimental Validation of Heat Transfer During Selective Laser Melting. In Proceedings of the ASME International Mechanical Engineering Congress and Exposition, Houston, TX, USA, 13–19 November 2015. [CrossRef]

50. Gusarov, A.V.; Yadroitsev, I.; Bertrand, P.; Smurov, I. Heat transfer modelling and stability analysis of selective laser melting. *Appl. Surf. Sci.* **2007**, *254*, 975–979. [CrossRef]

51. Li, C.; Wang, Y.; Zhan, H.; Han, T.; Han, B.; Zhao, W. Three-dimensional finite element analysis of temperatures and stresses in wide-band laser surface melting processing. *Mater. Des.* **2010**, *31*, 3366–3373. [CrossRef]

52. Fu, C.H.; Guo, Y.B. Three-Dimensional Temperature Gradient Mechanism in Selective Laser Melting of Ti-6Al-4V. *J. Manuf. Sci. Eng.* **2014**, *136*, 061004. [CrossRef]

53. Wen, S.F.; Li, S.; Wei., Q.S.; Yan, C.; Sheng, Z.; Shi, Y.S. Effects of molten pool boundaries on the mechanical properties of selective laser melted parts. *J. Mater. Process. Technol.* **2014**, *214*, 2660–2667. [CrossRef]

54. Heeling, T.; Cloots, M.; Wegener, K. Melt pool simulation for the evaluation of process parameters in selective laser melting. *Addit. Manuf.* **2017**, *14*, 116–125. [CrossRef]

55. Abe, F.; Osakada, K.; Shiomi, M.; Uematsu, K.; Matsumoto, M. The Manufacturing of Hard Tooling from Metallic Powders by Selective Laser Melting. *J. Mater. Process. Technol.* **2001**, *111*, 210–213. [CrossRef]

56. Wauthle, R.; Vrancken, B.; Beynaerts, B.; Jorissen, K.; Schrooten, J.; Kruth, J.-P.; van Humbeeck, J. Effects of build orientation and heat treatment on the microstructure and mechanical properties of selective laser melted Ti6Al4V lattice structures. *Addit. Manuf.* **2015**, *5*, 77–84. [CrossRef]

57. Mertens, R.; Clijsters, S.; Kempen, K.; Kruth, J.-P. Optimization of Scan Strategies in Selective Laser Melting of Aluminum Parts with Downfacing Areas. *J. Manuf. Sci. Eng.* **2014**, *136*, 061012. [CrossRef]

58. Vranken, B.; Thijs, L.; Kruth, J.-P.; van Humbeeck, J. Heat treatment of Ti6Al-4V produced by Selective Laser Melting: Microstructure and mechanical properties. *J. Alloy. Compd.* **2012**, *541*, 177–185. [CrossRef]

59. Dai, D.; Gu, D. Thermal behavior and densification mechanisms during selective laser melting of copper matrix composites: Simulation and experiments. *Mater. Des.* **2014**, *55*, 482–491. [CrossRef]
60. Wu, J.; Wang, L.; An, X. Numerical analysis of residual stress evolution of AlSi10Mg manufactured by selective laser melting. *Optik-Int. J. Light Electron Opt.* **2017**, *137*, 65–78. [CrossRef]
61. Fergani, O.; Berto, F.; Welo, T.; Liang, S.Y. Analytical modelling of residual stress in additive manufacturing. *Fatigue Fract. Eng. Mater. Struct.* **2016**. [CrossRef]
62. Vranken, B.; Cain, V.; Knutsen, R.; van Humbeeck, J. Residual stress via the contour method in compact tension specimens produced via selective laser melting. *Scripta Materialia* **2014**, *87*, 29–32. [CrossRef]
63. Zinovieva, O.; Zinoveiev, A.; Ploshikhin, V.; Romanova, V.; Balokhonov, R. Computational Study of the Mechanical Behavior of Steel Produced by Selective Laser Melting. *AIP Conf. Proc.* **2016**, *1783*, 020235. [CrossRef]
64. Liu, Y.; Yang, Y.; Wang, D. A study on the residual stress during selective laser melting (SLM) of metallic powder. *Int. J. Adv. Manuf. Technol.* **2016**, *87*, 647–656. [CrossRef]
65. Safronov, V.A.; Khmyrov, R.S.; Kotoban, D.V.; Gusarov, A.V. Distortions and Residual Stresses at Layer-by-Layer Additive Manufacturing by Fusion. *J. Manuf. Sci. Eng.* **2017**, *139*, 031017. [CrossRef]
66. Zaeh, M.F.; Branner, G. Investigation on residual stress and deformations in selective laser melting. *Prod. Eng. Res. Devel.* **2010**, *4*, 35–45. [CrossRef]
67. Gu, D.; He, B. Finite element simulation and experimental investigation of residual stresses in selective laser melted Ti-Ni shape memory alloy. *Comput. Mater. Sci.* **2016**, *117*, 221–232. [CrossRef]
68. Li, C.; Fu, C.H.; Guo, Y.B.; Fang, F.Z. A multiscale modeling approach for fast prediction of part distortion in selective laser melting. *J. Mater. Process. Technol.* **2016**, *229*, 703–712. [CrossRef]
69. Li., C.; Liu, J.F.; Guo, Y.B. Prediction of Residual Stress and Part Distortion in Selective Laser Melting. *Procedia CIRP* **2016**, *45*, 171–174. [CrossRef]
70. Parry, L.; Ashcroft, I.A.; Wildman, R.D. Understanding the effect of laser scan strategy on residual stress in selective laser melting through thermo-mechanical simulation. *Addit. Manuf.* **2016**, *12*, 1–15. [CrossRef]
71. Vastola, G.; Zhang, G.; Pei, Q.X.; Zhang, Y.-W. Controlling of residual stress in additive manufacturing of Ti6Al4V by finite element analysis. *Addit. Manuf.* **2016**, *12*, 231–239. [CrossRef]
72. Liu, F.; Lin, X.; Yang, G.; Sing, M.; Chen, J.; Huang, W. Microstructure and residual stress of laser rapid formed Inconel 718 nickel-base superalloy. *Opt. Laser Technol.* **2011**, *43*, 208–213. [CrossRef]
73. Toda-Caraballo, I.; Chao, J.; Lindgren, L.E.; Capdevila, C. Effect of residual stress on recrystallization behavior of mechanically alloyed steels. *Scripta Materialia* **2010**, *62*, 41–44. [CrossRef]
74. Chen, H.; Gu, D.; Dai, D.; Ma, C.; Xia, M. Microstructure and composition homogeneity, tensile property, and underlying thermal physical mechanism of selective laser melted tool steel parts. *Mater. Sci. Eng. A* **2017**, *682*, 279–289. [CrossRef]
75. Mertens, R.; Vrancken, B.; Holmstock, N.; Kinds, Y.; Kruth, J.-P.; van Humbeeck, J. Influence of powder bed preheating on microstructure and mechanical properties of H13 tool steel SLM parts. *Phys. Procedia* **2016**, *83*, 882–890. [CrossRef]
76. Vastola, G.; Zhang, G.; Pei, Q.X.; Zhang, Y.-W. Modeling the Microstructure Evolution During Additive Manufacturing of Ti-6Al-4V: A Comparison Between Electron Beam Melting and Selective Laser Melting. *JOM* **2016**, *68*, 1370–1375. [CrossRef]
77. Alyoshin, N.P.; Murashov, V.V.; Grigoryev, M.V.; Yevgenov, A.G.; Karachevtsev, F.N.; Shchipakov, N.A.; Vasilenko, S.A. Defects of heat-resistant alloys synthesized by the method of selective laser melting. *Inorg. Mater. Appl. Res.* **2017**, *8*, 27–31. [CrossRef]
78. Carter, L.N.; Martin, C.; Withers, P.J.; Attallah, M.M. The influence of the laser scan strategy on grain structure and cracking behavior in SLM powder-bed fabricated nickel superalloy. *J. Alloy. Compd.* **2014**, *615*, 338–347. [CrossRef]
79. Zhang, K.; Zhang, X.; Shang, X. Research on Cladding Process of Metal Powder During Laser Additive Manufacturing. *Appl. Mech. Mater.* **2013**, *380*, 4311–4314. [CrossRef]
80. Qian, Bo.; Shi, Y.; Wei, Q.; Wang, H. The helix scan strategy applied to the selective laser melting. *Int. J. Adv. Manuf. Technol.* **2012**, *63*, 631–640. [CrossRef]
81. Shiomi, M.; Osakada, K.; Nakamura, K.; Yamashita, T.; Abe, F. Residual Stress within Metallic Model Made by Selective Laser Melting Process. *CIRP Ann. Manuf. Technol.* **2004**, *53*, 195–198. [CrossRef]

82. Yasa, E.; Kruth, J.-P. Application of Laser Re-Melting on Selective Laser Melting Parts. *Adv. Prod. Eng. Manag.* **2011**, *6*, 259–270. Available online: https://lirias.kuleuven.be/bitstream/123456789/332611/2/APEM6-4_259-270.pdf (accessed on 11 April 2017).
83. Mumtaz, K.A.; Hopkinson, N. Selective Laser Melting of thin walled parts using pulse shaping. *J. Mater. Process. Technol.* **2010**, *210*, 279–287. [CrossRef]
84. Tolosa, I.; Garciandía, F.; Zubiri, F.; Zapirain, F.; Esnaola, A. Study of mechanical properties of AISI 316 stainless steel processed by "selective laser melting", following different manufacturing strategies. *Int. J. Adv. Manuf. Technol.* **2010**, *51*, 639–647. [CrossRef]
85. Brandl, E.; Heckenberger, U.; Holzinger, V.; Buchbinder, D. Additive manufactured AlSi10Mg samples using Selective Laser Melting (SLM): Microstructure, high cycle fatigue, and fracture behavior. *Mater. Des.* **2012**, *34*, 159–169. [CrossRef]
86. Edwards, P.; Ramulu, M. Fatigue performance evaluation of selective laser melted Ti-6Al-4V. *Mater. Sci. Eng. A* **2014**, *598*, 327–337. [CrossRef]
87. Guan, K.; Wang, Z.; Gao, M.; Li, X.; Zeng, X. Effects of processing parameters on tensile properties of selective laser melted 304 stainless steel. *Mater. Des.* **2013**, *50*, 581–586. [CrossRef]
88. Yadroitsev, I.; Smurov, I. Surface Morphology in Selective Laser Melting of Metal Powders. *Phys. Procedia* **2011**, *12*, 264–270. [CrossRef]
89. Yadroitsev, I.; Krakhmalev, P.; Yadroitsava, I.; Johansson, S.; Smurov, I. Energy input effect on morphology and microstructure of selective laser melting single track from metallic powder. *J. Mater. Process. Technol.* **2013**, *213*, 606–613. [CrossRef]
90. Cheng, B.; Shrestha, S.; Chou, L. Stress and deformation evaluations of scanning strategy effect in selective laser melting. *Addit. Manuf.* **2016**, *12*, 240–251. [CrossRef]
91. Xia, M.; Gu, D.; Yu, G.; Dai, D.; Chen, H.; Shi, Q. Influence of hatch spacing on heat and mass transfer, thermodynamics and laser processability during additive manufacturing of Inconel 718 alloy. *Int. J. Mach. Tools Manuf.* **2016**, *109*, 147–157. [CrossRef]
92. Lu, Y.; Gan, Y.; Lin, J.; Guo, S.; Wu, S.; Lin, J. Effect of laser speeds on the mechanical property and corrosion resistance of CoCrW alloy fabricated by SLM. *Rapid Prototyp. J.* **2017**, *23*, 28–33. [CrossRef]
93. Yu, G.; Gu, D.; Dai, D.; Xia, M.; Ma, C.; Chang, K. Influence of processing parameters on laser penetration depth and melting/re-melting densification during selective laser melting of aluminum alloy. *Appl. Phys. A* **2016**, *122*, 891. [CrossRef]
94. Jia, Q.; Gu, D. Selective laser melting additive manufactured Inconel 718 superalloy parts: High temperature oxidation property and its mechanisms. *Opt. Laser Technol.* **2014**, *62*, 161–171. [CrossRef]
95. Dadbakhsh, S.; Hao, L.; Sewell, N. Effect of selective laser melting layout on the quality of stainless steel parts. *Rapid Prototyp. J.* **2012**, *18*, 241–249. [CrossRef]
96. Ladewig, A.; Schlick, G.; Fisser, M.; Schulze, V.; Glatzel, U. Influence of the shielding gas flow on the removal of process by-products in the selective laser melting process. *Addit. Manuf.* **2016**, *10*, 1–9. [CrossRef]
97. Buchbinder, D.; Meiners, W.; Pirch, N.; Wissenbach, K.; Schrage, J. Investigation on reducing distortion by preheating during manufacture of aluminum components using selective laser melting. *J. Laser Appl.* **2014**, *26*, 012004. [CrossRef]
98. Craeghs, T.; Bechmann, F.; Berumen, S.; Kruth, J.-P. Feedback control of Layerwise Laser Melting using optical sensors. *Phys. Procedia* **2010**, *5*, 505–514. [CrossRef]
99. Craeghs, T.; Clijsters, S.; Yasa, E.; Bechmann, F.; Berumen, S.; Kruth, J.-P. Determination of geometrical factors in Layerwise Laser Melting using optical process monitoring. *Opt. Lasers Eng.* **2011**, *49*, 1440–1446. [CrossRef]
100. Craeghs, T.; Clijsters, S.; Yasa, E.; Kruth, J.-P. Online Quality Control of Selective Laser Melting. In Proceedings of the Solid Freeform Fabrication Symposium, Austin, TX, USA, 8–10 August 2011. Available online: https://sffsymposium.engr.utexas.edu/Manuscripts/2011/2011-17-Craeghs.pdf (accessed on 11 April 2017).
101. Craeghs, T.; Clijsters, S.; Kruth, J.-P.; Bechmann, F.; Ebert, M.-C. Detection of process failures in Layerwise Laser Melting with optical process monitoring. *Phys. Procedia* **2012**, *39*, 753–759. [CrossRef]
102. Devesse, W.; de Baere, D.; Hinderdael, M.; Guillaume, P. Hardware-in-the-loop control of additive manufacturing processes using temperature feedback. *J. Laser Appl.* **2016**, *28*, 022302. [CrossRef]
103. Pacurar, R.; Balc, N.; Prem, F. Research on how to improve the accuracy of the SLM metallic parts. In AIP Conference Proceedings, Belfast, UK, 27–29 April 2011.

104. Casalino, G.; Campanelli, S.L.; Contuzzi, N.; Ludovico, A.D. Experimental investigation and statistical optimization of the selective laser melting process of a maraging steel. *Opt. Laser Technol.* **2015**, *65*, 151–158. [CrossRef]
105. Aboutaleb, A.M.; Bain, L.; Elwany, A.; Shamsaei, N.; Thompson, S.M.; Tapia, G. Accelerated process optimization for laser-based additive manufacturing by leveraging similar prior studies. *IISE Trans.* **2017**, *49*, 31–44. [CrossRef]
106. Zhao, X.; Lin, X.; Chen, J.; Xue, L.; Huang, W. The effect of hot isostatic pressing on crack healing, microstructure, mechanical properties of Rene88DT superalloy prepared by laser solid forming. *Mater. Sci. Eng. A* **2009**, *504*, 129–134. [CrossRef]
107. Campanelli, S.L.; Contuzzi, N.; Ludovico, A.D.; Caiazzo, F.; Cardaropoli, F.; Sergi, V. Manufacturing and Characterization of Ti6Al4V Lattice Components Manufactured by Selective Laser Melting. *Materials* **2014**, *7*, 4803–4822. [CrossRef]
108. AlMongour, B.; Grzesiak, D.; Yang, J.-M. Selective laser melting of TiB2/H13 steel nanocomposites: Influence of hot isostatic pressing post-treatment. *J. Mater. Process. Technol.* **2017**, *244*, 344–353. [CrossRef]
109. AlMongour, B.; Grzesiak, D.; Yang, J.-M. Selective laser melting of TiB2/316L stainless steel composites: The roles of powder preparation and hot isostatic pressing post-treatment. *Powder Technol.* **2017**, *309*, 37–48. [CrossRef]
110. AlMongour, B.; Yang, J.-M. Understanding the deformation behavior of 17-4 precipitate hardenable stainless steel produced by direct metal laser sintering using micropillar compression and TEM. *Int. J. Adv. Manuf. Technol.* **2017**, *90*, 119–126. [CrossRef]
111. Kreitcberg, A.; Brailovski, V.; Turenne, S. Effect of heat treatment and hot isostatic pressing on the microstructure and mechanical properties of Inconel 625 alloy processed by laser powder bed fusion. *Mater. Science Eng. A* **2017**, *689*, 1–10. [CrossRef]
112. Li, W.; Li, S.; Liu, J.; Zhang, A.; Zhou, Y.; Wei, Q.; Yan, C.; Shi, Y. Effect of heat treatment on AlSi10Mg alloy fabricated by selective laser melting: Microstructure evolution, mechanical properties and fracture mechanism. *Mater. Sci. Eng. A* **2016**, *663*, 116–125. [CrossRef]
113. Tillmann, W.; Schaak, C.; Nellesen, J.; Schaper, M.; Aydinöz, M.E.; Hoyer, K.-P. Hot isostatic pressing of IN718 components manufactured by selective laser melting. *Addit. Manuf.* **2017**, *13*, 93–102. [CrossRef]
114. Tucho, W.M.; Cuvillier, P.; Sjolyst-Kverneland, A.; Hanson, V. Microstructure and hardness studies of Inconel 718 manufactured by selective laser melting before and after solution heat treatment. *Mater. Sci. Eng. A* **2017**, *689*, 220–232. [CrossRef]
115. Fiocchi, J.; Tuissi, A.; Bassani, P.; Biffi, C.A. Low temperature annealing dedicated to AlSi10Mg selective laser melting products. *J. Alloy. Compd.* **2017**, *695*, 3402–3409. [CrossRef]
116. Song, B.; Dong, S.; Liu, Q.; Liao, H.; Coddet, C. Vacuum heat treatment of iron parts produced by selective laser melting: Microstructure, residual stress, and tensile behavior. *Mater. Des.* **2014**, *54*, 727–733. [CrossRef]
117. AlMangour, B.; Yang, J.-M. Improving the surface quality and mechanical properties by shot-peening of 17-4 stainless steel fabricated by additive manufacturing. *Mater. Des.* **2016**, *110*, 914–924. [CrossRef]
118. Aleshin, N.P.; Gregor'ev, M.V.; Murashov, V.V.; Krasnov, I.S.; Krupnina, O.A.; Smorodinskii, Y.G. Assessing the Results of Ultrasonic Testing of Additive Manufactured Parts with Alternative Methods. *Russ. J. Nondestruct. Test.* **2016**, *52*, 691–696. [CrossRef]
119. Kleszczynski, S.; zur Jacobsmühlen, J.; Witt, G. Error Detection in Laser Beam Melting Systems by High Resolution Imaging. In Proceedings of the Twenty Third Annual International Solid Freeform Fabrication Symposium, Austin, TX, USA, 6–8 August 2012. Available online: https://sffsymposium.engr.utexas.edu/Manuscripts/2012/2012-74-Kleszczynski.pdf (accessed on 11 April 2017).
120. Clijsters, S.; Craeghs, T.; Buls, S.; Kempen, K.; Kruth, J.-P. In situ quality control of the selective laser melting process using a high-speed, real-time melt pool monitoring system. *Int. J. Adv. Technol.* **2014**, *75*, 1089–1101. [CrossRef]
121. Chivel, Y. Optical in-process temperature monitoring of selective laser melting. *Phys. Procedia* **2013**, *41*, 904–910. [CrossRef]
122. Grasso, M.; Laguzza, V.; Semeraro, Q.; Colosimo, B.M. In-Process Monitoring of Selective Laser Melting: Spatial Detection of Defects Via Image Data Analysis. *J. Manuf. Sci. Eng.* **2017**, *139*, 051001. [CrossRef]

123. Hirsh, M.; Patel, R.; Li, W.; Guan, G.; Leach, R.K.; Sharples, S.D.; Clare, A.T. Assessing the capability of in-situ nondestructive analysis during layer based additive manufacturing. *Addit. Manuf.* **2017**, *13*, 135–142. [CrossRef]

124. Kanko, J.A.; Sibley, A.P.; Fraser, J.M. In situ morphology-based defect detection of selective laser melting through inline coherent imaging. *J. Mater. Process. Technol.* **2016**, *231*, 488–500. [CrossRef]

125. Lott, P.; Schleifenbaum, H.; Meiners, W.; Wissenbach, K.; Hinke, C.; Bültmann, J. Design of an Optical system for the In Situ Process Monitoring of Selective Laser Melting (SLM). *Phys. Procedia* **2011**, *12*, 683–690. [CrossRef]

126. Rodriguez, E.; Mireles, J.; Terrazas, C.A.; Espalin, D.; Perez, M.A.; Wicker, R.B. Approximation of absolute surface temperature measurements of powder bed fusion additive manufacturing technology using in situ infrared thermography. *Addit. Manuf.* **2015**, *5*, 31–39. [CrossRef]

127. Smurov, I.Y.; Dubenskaya, M.A.; Zhirnov, I.V.; Teleshevskii, V.I. Determination of the True Temperature During Selective Laser Melting of Metal Powders Based on Measurements with an Infrared Camera. *Meas. Tech.* **2016**, *59*, 971–974. [CrossRef]

128. Casavola, C.; Campanelli, S.L.; Pappalettere, C. Preliminary investigation on distribution of residual stresses generated by the selective laser melting process. *J. Strain Anal.* **2009**, *44*, 93–104. [CrossRef]

129. Yadroitsava, I.; Yadroitsev, I. Residual Stress in Metal Specimens Produced by Direct Metal Laser Sintering. Available online: https://sffsymposium.engr.utexas.edu/sites/default/files/2015/2015-49-Yadroitsev.pdf (accessed on 11 April 2017).

130. Dunbar, A.J.; Denlinger, E.R.; Heigel, J.; Michaleris, P.; Guerrier, P.; Mertukanitz, R.; Simpson, T.W. Development of experimental method for in situ distortion and temperature measurements during the laser powder bed fusion additive manufacturing process. *Addit. Manuf.* **2016**, *12*, 25–30. [CrossRef]

131. Havermann, D.; Mathew, J.; MacPherson, W.N.; Hand, D.P.; Maier, R.R.J. Measuring Residual Stresses in metallic components manufactured with Fibre Bragg Gratings embedded by Selective Laser Melting. In Proceedings of the International Conference on Optical Fibre Sensors (OFS24), Curitiba, Brazil, 28 September 2015. [CrossRef]

132. Yadroitsava, I.; Grewar, S.; Hattingh, D.; Yadroitsev, I. Residual Stress in SLM Ti6Al4V Alloy Specimens. *Mater. Sci. Forum* **2015**, *828*, 305–310. [CrossRef]

133. Protasov, C.E.; Safronov, V.A.; Kotoban, D.V.; Gusarov, A.V. Experimental study of residual stresses in metal parts obtained by selective laser melting. *Phys. Procedia* **2016**, *83*, 825–832. [CrossRef]

134. Sundar, R.; Hedaoo, P.; Ranganathan, K.; Bindra, K.S.; Oak, S.M. Application of Meshes to Extract the Fabricated Objects in Selective Laser Melting. *Mater. Manuf. Process.* **2014**, *29*, 429–433. [CrossRef]

135. Maliaris, G.; Sarafis, I.T.; Lazaridis, T.; Varoutoglou, A.; Tsakataras, G. Random lattice structures. Modeling, manufacture and FEA of their mechanical response. *Mater. Sci. Eng.* **2016**, *161*, 012045. [CrossRef]

136. Strano, G.; Hao, L.; Everson, R.M.; Evans, K.E. A new approach to the design and optimization of support structures in additive manufacturing. *Int. J. Adv. Manuf. Technol.* **2013**, *66*, 1247–1254. [CrossRef]

137. Wang, D.; Yang, Y.; Yi, Z.; Su, X. Research on the fabricating quality optimization of the overhanging surface in SLM process. *Int. J. Adv. Manuf. Technol.* **2013**, *65*, 1471–1484. [CrossRef]

138. Wang, D.; Mai, S.; Xiao, D.; Yang, Y. Surface quality of the curved overhanging structure manufactured from 316-L stainless steel by SLM. *Int. J. Adv. Manuf. Technol.* **2016**, *86*, 781–792. [CrossRef]

139. Cloots, M.; Spierings, A.B.; Wegener, K. Assessing new support minimizing strategies for the additive manufacturing technology SLM. In Proceedings of the 24th International SFF Symposium—An Additive Manufacturing Conference, Austin, TX, USA, 12–14 August 2013. Available online: https://sffsymposium. engr.utexas.edu/Manuscripts/2013/2013-50-Cloots.pdf (accessed on 11 April 2017).

140. Fox, J.C.; Moylan, S.P.; Lane, B.M. Effect of process parameters on the surface roughness of overhanging structures in laser powder bed fusion additive manufacturing. *Procedia CIRP* **2016**, *45*, 131–134. [CrossRef]

141. Kruth, J.-P.; Vandenbrouke, B.; van Vaerenbergh, J.; Mercelis, P. Benchmarking of Different SLS/SLM Processes as Rapid Manufacturing Techniques. In Proceedings of the International Conference on Polymers and Moulds Innovations, Gent, Belgium, 20–23 April 2005. Available online: https://core.ac.uk/download/pdf/11459701.pdf (accessed on 11 April 2017).

142. Papalambros, P.Y.; Wilde, D.J. *Principles of Optimal Design*, 2nd ed.; Cambridge University Press: Cambridge, UK, 2000; pp. 337–350.

MDPI AG

St. Alban-Anlage 66

4052 Basel, Switzerland

Tel. +41 61 683 77 34

Fax +41 61 302 89 18

http://www.mdpi.com

Technologies Editorial Office

E-mail: technologies@mdpi.com

http://www.mdpi.com/journal/technologies